AF576022

Diet Quality and Risk of Cardiometabolic and Diabetes

Diet Quality and Risk of Cardiometabolic and Diabetes

Editor

Giuseppe Della Pepa

Basel • Beijing • Wuhan • Barcelona • Belgrade • Novi Sad • Cluj • Manchester

Editor
Giuseppe Della Pepa
National Research
Council-CNR
Pisa, Italy

Editorial Office
MDPI
St. Alban-Anlage 66
4052 Basel, Switzerland

This is a reprint of articles from the Special Issue published online in the open access journal *Nutrients* (ISSN 2072-6643) (available at: https://www.mdpi.com/journal/nutrients/special_issues/Diet_Cardiometabolic).

For citation purposes, cite each article independently as indicated on the article page online and as indicated below:

Lastname, A.A.; Lastname, B.B. Article Title. *Journal Name* **Year**, *Volume Number*, Page Range.

ISBN 978-3-0365-8774-5 (Hbk)
ISBN 978-3-0365-8775-2 (PDF)
doi.org/10.3390/books978-3-0365-8775-2

© 2023 by the authors. Articles in this book are Open Access and distributed under the Creative Commons Attribution (CC BY) license. The book as a whole is distributed by MDPI under the terms and conditions of the Creative Commons Attribution-NonCommercial-NoDerivs (CC BY-NC-ND) license.

Contents

Preface

Cardiovascular diseases (CVD) are the leading causes of global disease burden and disease-related mortality. The evolution of numerous cardiometabolic risk factors and type 2 diabetes (T2D) related to CVD is driven by visceral obesity, and behavioral lifestyle weight loss therapies are crucial preventative measures that can counteract these metabolic changes.

The long-term maintenance of weight loss following low-calorie diets poses a significant challenge, despite the overwhelming evidence that suggests that the greater the body weight loss, the greater the preventive effect on cardiometabolic risk factors or diabetes.

Greater adherence to suggested dietary patterns and/or consumption of dietary patterns linked to a lower risk of cardiometabolic diseases and other chronic diseases are two characteristics of higher diet quality. There is strong evidence from prospective cohort studies showing that higher food quality is linked to a decreased risk of CVD and T2D.

Therefore, current dietary advice for overall health and cardiometabolic prevention and management places a strong emphasis on maintaining a good dietary pattern over the course of a person's lifetime. High diet quality is a distinguishing characteristic of all recommended dietary patterns for general health and cardiometabolic prevention and treatment, despite the fact that there are minor variations in the precise food- and nutrient-based recommendations made by authorized organizations. A healthy diet is generally agreed to be one that is high in fruits, vegetables, whole grains, legumes, nuts, seeds, low-fat dairy, and foods with lean protein and low levels of saturated fat, added sugar, and sodium. Therefore, rather than arguing over specific food, food group, or nutrient recommendations, attention should be paid to the overall diet, areas of relative agreement about core healthy foods, and the creation of techniques that might successfully nudge people toward improved diet quality.

Changes in diet composition acting on nutrient quality independently of changes in energy intake may be effective in cardiometabolic and T2D risk prevention, offering a more feasible and safe alternative treatment to energy restriction.

This book collates articles summarizing recent evidence on "Diet Quality and Risk of Cardiometabolic and Diabetes".

The book explores more specifically the impact of diet quality in terms of micro- or macronutrient composition, beyond the effect of diet restriction, on the prevention of cardiometabolic and diabetes risk as well as diabetes management. Personalized, quality dietary interventions for cardiometabolic health and diabetes prevention, as well as possible underlying mechanisms, will also be addressed in this book.

This book is relevant to any student or practitioner interested in how diet influences our health in the fields of nutrition, dietetics, medicine, and public health.

In the end, we would like to take this opportunity to express our most profound appreciation to the MDPI Book staff, the editorial team of Nutrients, the assistant editor of this Special Issue, the talented authors, and hardworking and professional reviewers.

Giuseppe Della Pepa
Editor

Article

Diet Quality Is Associated with Glucose Regulation in a Cohort of Young Adults

Elizabeth Costello [1,*], Jesse Goodrich [1], William B. Patterson [2], Sarah Rock [1], Yiping Li [1], Brittney Baumert [1], Frank Gilliland [1], Michael I. Goran [1,3], Zhanghua Chen [1], Tanya L. Alderete [2], David V. Conti [1] and Leda Chatzi [1]

1 Department of Population and Public Health Sciences, University of Southern California, Los Angeles, CA 90032, USA
2 Department of Integrative Physiology, University of Colorado Boulder, Boulder, CO 80309, USA
3 Department of Pediatrics, The Saban Research Institute, Children's Hospital Los Angeles, Los Angeles, CA 90027, USA
* Correspondence: eecostel@usc.edu

Abstract: Young-onset type 2 diabetes and prediabetes is a growing epidemic. Poor diet is a known risk factor for T2D in older adults, but the contribution of diet to risk factors for T2D is not well-described in youth. Our objective was to examine the relationship of diet quality with prediabetes, glucose regulation, and adiposity in young adults. A cohort of young adults (n = 155, age 17–22) was examined between 2014–2018, and 89 underwent a follow-up visit from 2020–2022. At each visit, participants completed diet and body composition assessments and an oral glucose tolerance test. Adherence to four dietary patterns was assessed: Dietary Approaches to Stop Hypertension (DASH), Healthy Eating Index (HEI), Mediterranean diet, and Diet Inflammatory Index (DII). Regression analyses were used to determine adjusted associations of diet with risk for prediabetes and adiposity. Each one-point increase in DASH or HEI scores between visits reduced the risk for prediabetes at follow-up by 64% (OR, 95% CI: 0.36, 0.17–0.68) and 9% (OR, 95% CI: 0.91, 0.85–0.96), respectively. The DASH diet was inversely associated with adiposity, while DII was positively associated with adiposity. In summary, positive changes in HEI and DASH scores were associated with reduced risk for prediabetes in young adults.

Keywords: diet quality; dietary patterns; type 2 diabetes; prediabetes; obesity; body composition; young adults

Citation: Costello, E.; Goodrich, J.; Patterson, W.B.; Rock, S.; Li, Y.; Baumert, B.; Gilliland, F.; Goran, M.I.; Chen, Z.; Alderete, T.L.; et al. Diet Quality Is Associated with Glucose Regulation in a Cohort of Young Adults. *Nutrients* **2022**, *14*, 3734. https://doi.org/10.3390/nu14183734

Academic Editor: Giuseppe Della Pepa

Received: 18 August 2022
Accepted: 8 September 2022
Published: 10 September 2022

Publisher's Note: MDPI stays neutral with regard to jurisdictional claims in published maps and institutional affiliations.

Copyright: © 2022 by the authors. Licensee MDPI, Basel, Switzerland. This article is an open access article distributed under the terms and conditions of the Creative Commons Attribution (CC BY) license (https://creativecommons.org/licenses/by/4.0/).

1. Introduction

The prevalence of prediabetes, a condition where blood glucose levels are elevated but below diagnostic cut-offs for type 2 diabetes (T2D) [1], is increasing in adolescents and young adults in the United States (U.S.) [2,3]. Prediabetes greatly increases the risk for T2D [4]; therefore, T2D incidence is also increasing in the U.S., following a similar trend [5]. This trend is of considerable concern because T2D is often more aggressive in youth than in older adults and is associated with higher rates of complications, more comorbidities, and higher mortality risk [6,7]. Disparities also exist in T2D risk, with Hispanics and other racial or ethnic minorities at higher risk compared to non-Hispanic Whites [5,7,8]. Lifestyle is a source of modifiable risk factors frequently targeted in preventive measures [1,9], of which diet is especially important.

Depending on quality, diet may be either protective against or a risk factor for prediabetes and T2D [10–12]. Healthy dietary patterns high in fruits, vegetables, and whole grains and low in sodium, saturated fat, and added sugars are associated with reduced risk for prediabetes and T2D [10,13–15]. In middle-aged and older adults, adherence to healthy eating patterns such as the Mediterranean diet, Dietary Approaches to Stop Hypertension (DASH) diet, and federal dietary recommendations reduces the risk for T2D [13,14,16]. The Dietary Inflammatory Index (DII), an alternative dietary pattern that quantifies the inflammatory effects of dietary intake, is linked with prediabetes and T2D, where more pro-inflammatory diets are associated with increased risk [17,18]. However, most studies evaluating the relationship between diet and T2D risk have been conducted in middle-aged or older adults or only incidentally included young adults [11,19,20]. Less is understood about the impact of diet quality or dietary changes on T2D risk in young adulthood.

Few prospective studies have examined the relationships between the DASH diet, Mediterranean diet, or other dietary patterns and T2D in youth [21–23]. Findings in children and adolescents suggest that increased adherence to the DASH diet may improve cardiovascular and metabolic risk factors [21] and that weight control is critical in T2D prevention [22,24]. Limited studies exist on the development of T2D in young adults [25–28] though this life stage may represent a critical window for behavior change and diabetes prevention, as young people transition from their adolescent years into independent adulthood [29].

The purpose of this study was to examine the relationship between diet quality and risk for T2D in a cohort of primarily Hispanic young adults. Participants were evaluated for glucose dysregulation and diet quality at age 17–22 and again after approximately four years. Glucose regulation was assessed using hemoglobin A1c (HbA1c) and 2-h oral glucose tolerance tests (OGTTs). We hypothesize that higher diet quality will be protective against glucose dysregulation and that improvement in diet quality between visits will reduce the risk for prediabetes and type 2 diabetes.

2. Materials and Methods

2.1. Cohort

Between 2014 and 2018, a subset of 158 participants between 17 and 22 years old were recruited from the Children's Health Study (CHS) [30] for the Meta-AIR study [31]. Subjects were selected if they had overweight or obesity in early adolescence, had not been diagnosed with type 1 or type 2 diabetes, had no medical conditions, and were taking no medications that affect glucose metabolism [31]. Between January 2020 and March 2022, 140 of these participants were invited to participate in a follow-up visit. All but 7 participants underwent follow-up testing during the COVID-19 pandemic. All study visits were completed at the Diabetes and Obesity Research Institute at the University of Southern California. This study was approved by the University of Southern California Institutional Review Board. Written informed consent was obtained from participants at both baseline and follow-up visits or by participants and their guardians for those under age 18 at baseline.

Of the 158 participants at baseline, 155 had diet data and data for at least one outcome. Eighty-six of these participated in the follow-up (Figure S1). An additional three CHS participants without baseline data completed the follow-up visit.

2.2. Glucose Outcomes

A 2-h oral glucose tolerance test (OGTT) was performed at each visit, using a glucose load of 1.75 g of glucose per kg of body mass (max 75 g). Blood was sampled before the glucose challenge and at 30-, 60-, 90- and 120-min post-challenge. Glucose concentrations at each time point were measured in plasma. Fasting glucose was also measured using a glucometer before the glucose challenge, and the OGTT was not completed if participants had a fasting value greater than 126 mg/dL. Hemoglobin A1c (HbA1c) was measured in fasting whole blood samples. Glucose area under the curve (AUC) was calculated from the five glucose measurements using the trapezoidal method [32].

Prediabetes and type 2 diabetes were based on clinical cutoffs for HbA1c, fasting plasma glucose, or 2-h plasma glucose [33]. Participants having either HbA1c values of 6.5% or higher, fasting glucose of 126 mg/dL or higher, or 2-h glucose of 200 mg/dL or higher were considered to have type 2 diabetes, while those with HbA1c between 5.7% and 6.4%, fasting glucose between 100 mg/dL and 125 mg/dL, or 2-h glucose between 140 mg/dL and 199 mg/dL were categorized as prediabetic.

2.3. Adiposity Outcomes

Weight and height were measured at each visit, and BMI calculated as kg per meter squared (kg/m^2). Body composition was assessed using dual-energy X-ray absorptiometry (DEXA) whole body scans. Baseline scans were performed on either a Hologic QDR 4500W or Horizon W machine at baseline, while all follow-up scans were performed on the Horizon W. Body composition measures included total body fat percentage, fat mass to height ratio, fat free mass index (FFMI, kg/m^2), android to gynoid ratio, trunk to leg ratio, trunk to limb ratio, and visceral adipose tissue (VAT, in^3). Only body fat percentage and fat mass to height ratio were measured on the QDR 4500W machine.

2.4. Diet Assessment

At each visit, participants were asked to complete two 24-h dietary recalls on non-consecutive days: one weekday and one weekend day. Baseline recalls were completed by trained interviewers using the Nutritional Data System for Research (NDSR) software version 2014, developed by the Nutrition Coordinating Center (University of Minnesota, Minneapolis, MN, USA) [34], while follow-up recalls used the Automated Self-Administered 24-h (ASA24) Dietary Assessment Tool, version (2018), developed by the National Cancer Institute, Bethesda, MD, USA [35]. An average of the values from both days was calculated for each diet component. At baseline, 16 participants (10.3%) completed only one recall, and 9 (10.2%) completed only one recall at follow-up. If a participant did not complete both recalls, values from the single recall were used.

Four diet indices were calculated from the recall data at both the baseline and follow-up visits: the 2015 Healthy Eating Index (HEI), DASH score, Mediterranean Diet Score (MDS), and DII. The HEI ranges from 0–100, is based on adherence to the United States Department of Agriculture (USDA) 2015 Dietary Guidelines [36] and contains the following thirteen elements standardized to calorie intake: total fruit, whole fruit, total vegetables, greens and beans, whole grains, refined grains, dairy, total protein foods, seafood and plant proteins, mono- and polyunsaturated fatty acids, saturated fats, sodium, and added sugars. The DASH scoring method follows the calculation proposed by Mellen et al. [37], using nutrient goals for DASH diet adherence. This DASH score ranges from 0 to 8 and includes the following elements standardized to calorie intake: protein, fiber, magnesium, calcium, potassium, total fat, saturated fat, cholesterol, and sodium. One point was assigned if the nutrient goal was met, and half of a point was assigned if an intermediate nutrient goal was met. The MDS was calculated using the method developed by Trichopoulou et al. [38], which ranges from 0 to 9 with ten components: vegetables, legumes, fruits and nuts, dairy, cereals, meat, poultry, fish, alcohol, and ratio of mono- to saturated fats. For each component, one point was assigned for exceeding the sex-specific median. The DII was adapted from Shivappa et al. [39], with negative values indicating an anti-inflammatory diet and positive values indicating a pro-inflammatory diet. For each element, a centered percentile was calculated by comparing the reported intake to a global mean and standard deviation of intake. This centered percentile was multiplied by the element's overall inflammatory effect score, and the scores for all elements were summed to produce the DII score. Twenty-eight of the forty-five elements from Shivappa et al. [39] were included: alcohol, beta-carotene, caffeine, carbohydrates, cholesterol, calories, total fat, fiber, iron, magnesium, folic acid, mono- and polyunsaturated fatty acids, omega-3 and omega-6 fatty acids, protein, saturated fat, selenium, zinc, and vitamins A, B1 (thiamin), B2 (riboflavin), B3 (niacin), C, D, and E. The remaining elements were excluded because they are not specifically captured by the recalls systems used in this study. Trans fats were banned by the United States Food and Drug Administration in 2015, with a 2018 deadline for implementation [40], and were excluded from the DII calculation in the follow-up visit.

2.5. Covariates

Questionnaires were administered to collect sociodemographic information, including age, sex, race and ethnicity, physical activity, and parental education. Ethnicity was categorized as non-Hispanic White, Hispanic, or Other. Parental education was categorized as "Did not complete high school", "Completed high school", "Completed more than high school", or "Don't know". At baseline, physical activity was assessed as a binary variable, where participants responded yes or no to the question "Do you exercise?". At follow-up, physical activity was assessed using the International Physical Activity Questionnaire Short Form [41], and metabolic equivalent of task (MET) minutes were calculated according to the scoring guidelines. Participants were considered to have "high" physical activity if they met either of the following criteria: (1) reported vigorous physical activity (VPA) three or more days per week and 1500 or more MET min per week or (2) seven days of any combination of VPA, moderate physical activity (MPA), and walking for at least 3000 MET min. Participants were considered to have "moderate" physical activity if they (1) reported at least 3 days of VPA where the activity lasted at least 30 min or (2) five or more days of MPA or walking where the activity lasted at least 30 min or (3) five or more days of some combination of VPA, MPA, and walking for at least 600 MET min. Participants were categorized as having "low" physical activity if they did not meet any of these criteria.

2.6. Statistical Analysis

Descriptive statistics were calculated for all outcomes and exposures. Pearson's correlations were calculated between the four diet scores at each visit separately and between time points. Independent two-sample t-tests, chi-square tests, or Fisher's exact tests were used to test for differences in participant demographics between the baseline cohort and follow-up cohort. Paired t-tests or McNemar–Bowker tests were used to test for differences in exposures and outcomes between visits. Due to the small numbers of participants with values meeting the diagnostic criteria for type 2 diabetes, prediabetes and diabetes were combined into one category (prediabetes/T2D) for analysis. Primary outcomes of interest were those related to glucose regulation: prediabetes/T2D, fasting glucose, 2-h glucose, glucose AUC, and HbA1c. Body composition measurements were secondary outcomes: BMI, body fat percent, FFMI, fat mass to height ratio, android to gynoid ratio, trunk to leg ratio, trunk to limb ratio, and VAT.

Cross-sectional analyses were performed for both baseline and follow-up visits, using multivariable linear regression for continuous outcomes and logistic regression for prediabetes/T2D. For longitudinal analyses, change in diet indices from baseline to follow-up was modeled against change in outcome using linear regression for continuous outcomes, or against diabetes at follow-up using logistic regression. Longitudinal models also adjusted for baseline diet score. Beta coefficients for exposures were scaled to one standard deviation (SD) of the exposure to account for the differing scales.

All analyses included the following covariates: age, ethnicity, physical activity, energy intake, and parental education. Because these factors were not accounted for in the scoring system, analyses with HEI, DASH, and DII scores additionally controlled for sex, and analyses with MDS additionally controlled for energy intake. BMI and body fat percent were presumed to be on the causal pathway between diet and prediabetes and T2D and were not included as covariates in the main analyses to avoid overadjustment [42].

2.7. Sensitivity Analyses

For all diet indices and glucose outcomes, two additional analyses were performed. The first did not include physical activity in as a covariate to determine if it had the potential to confound the relationship between diet and glucose regulation and if it was necessary to control for this variable in the main analysis. The second analysis controlled for body fat percent. Though we expect that body fat (or BMI) is on the causal pathway between diet and T2D, we included it as a covariate to examine the possibility that body fat mediates the relationship between diet and T2D.

We also performed additional logistic regression analyses to examine the association between each adiposity measure and risk for prediabetes/T2D at each visit and to examine the associations between changes in these measures between visits and risk for prediabetes/T2D at the follow-up visit. Models were adjusted for age, sex, ethnicity, parental education, energy intake, and physical activity as in the main analyses.

3. Results

Average length of follow-up was 4.1 years (SD = 1.1 years). There were no differences in participant demographics at each visit (Table 1). HEI, DASH, and DII scores significantly decreased from baseline to follow-up (Table 2), and mean fasting glucose and glucose AUC increased (Table 3). Mean BMI and body fat percentage also increased between visits (Table 4).

Table 1. Descriptive statistics for participant demographics at baseline and follow-up.

	Baseline (n = 155)	**Follow-Up (n = 88)** [1]	**Baseline vs. Follow-Up p-Value** [2]
Age (years), Mean (SD)	19.7 (1.2)	24.1 (0.8)	-
Sex, n (%)			
Female	71 (45.8)	46 (52.3)	0.40
Male	84 (54.2)	42 (47.7)	
Ethnicity, n (%)			
Hispanic/Latino	94 (60.6)	50 (56.8)	0.60
Non-Hispanic White	52 (33.5)	30 (34.1)	
Other	9 (5.8)	8 (9.1)	
Parental Education, n (%)			
Did not complete high school	31 (20.0)	15 (17.0)	0.76
Completed high school	23 (14.8)	12 (13.6)	
More than high school	96 (61.9)	56 (63.6)	
Don't know	5 (3.2)	5 (5.7)	
Exercise [3], n (%)			
Yes	118 (76.1)	-	-
No	37 (23.9)		
Physical Activity Category, n (%)			
High	-	50 (56.8)	-
Moderate		21 (23.9)	
Low		16 (18.2)	
Missing, n (%)		1 (1.1)	

[1] Includes three participants who did not complete the baseline visit. [2] p-values calculated using chi-Square or Fisher's exact tests. [3] Response to the question "Do you exercise?". SD: standard deviation.

Table 2. Descriptive statistics for diet at baseline, follow-up, and change between visits.

	Baseline (n = 155)	**Follow-Up (n = 88)**	**Change between Baseline and Follow-Up (n = 85)** [1]	**Baseline vs. Follow-Up p-Value** [2]
HEI, Mean (SD) Range: 0–100	52.7 (13.0)	49.7 (12.5)	−4.9 (13.2)	<0.001
MDS, Mean (SD) Range: 0–9	5.03 (1.23)	4.92 (1.53)	−0.22 (1.79)	0.25
DASH, Mean (SD) Range: 0–8	2.26 (1.51)	1.74 (1.31)	−0.45 (1.53)	0.009
DII, Mean (SD)	0.81 (1.56)	0.29 (2.05)	−0.44 (1.98)	0.044
Energy (kcal), Mean (SD)	2053 (630)	2223 (773)	158 (792)	0.070

[1] Three additional CHS participants participated in the second visit without having completed the first. [2] p-values calculated using paired t-tests. Abbreviations: HEI: Healthy Eating Index—2015; MDS: Mediterranean Diet Score; DASH: Dietary Approaches to Stop Hypertension; DII: Dietary Inflammatory Index.

Table 3. Descriptive statistics for glucose outcomes at baseline, follow-up, and change between visits.

	Baseline (n = 155)	Follow-Up (n = 88)	Change between Baseline and Follow-Up (n = 85) [1]	Baseline vs. Follow-Up p-Value [2]
Fasting Glucose, Mean (SD) Missing: n (%)	91. (14) 1 (0.6)	95 (16) 1 (1.1)	5 (15) 1 (1.2%)	0.003
2-h Glucose, Mean (SD) Missing: n (%)	123 (37) 1 (0.6)	122 (35) 4 (4.5)	3 (32) 4 (4.7)	0.39
HbA1c, Mean (SD) Missing: n (%)	5.25 (0.53) 1 (0.6)	5.26 (0.51)	0.042 (0.46)	0.35
Glucose AUC, Mean (SD) Missing: n (%)	267 (59) 1 (0.6)	269 (44) 6 (6.8)	11 (40) 6 (7.1)	0.023
Diabetes, n (%) No Diabetes Prediabetes Type 2 Diabetes Missing	 109 (70.3) 42 (27.1) 3 (1.9) 1 (0.6)	 54 (61.4) 30 (34.1) 4 (4.5)		0.17

[1] Three additional CHS participants participated in the second visit without having completed the first. [2] p-values calculated using paired t-tests for continuous variables and McNemar–Bowker test for diabetes categories. Abbreviations: SD: standard deviation; HbA1c: hemoglobin A1c; AUC: area under the curve.

Table 4. Descriptive statistics for body composition at baseline, follow-up, and change between visits.

	Baseline (n = 155)	Follow-Up (n = 88)	Change between Baseline and Follow-Up (n = 85) [1]	Baseline vs. Follow-Up p-Value [2,3]
BMI Category, n (%) Normal Weight Overweight Obese	 24 (15.5) 73 (47.1) 58 (37.4)	 12 (13.6) 34 (38.6) 42 (47.7)		0.47
BMI (kg/m^2), Mean (SD)	29.9 (5.1)	31.7 (7.0)	1.8 (4.3)	<0.001
Body Fat %, Mean (SD) Missing: n (%)	34.8 (8.6) -	38.5 (8.3) 2 (2.3)	3.1 (4.7) 2 (2.4)	<0.001
FFMI (kg/m^2), Mean (SD) Missing: n (%)	18.5 (2.5) -	17.7 (2.9) 2 (2.3)	−0.6 (1.5) 2 (2.4)	0.001
Fat Mass:Height Ratio, Mean (SD) Missing: n (%)	10.8 (4.3) 98 (63.2)	12.2 (4.7) 2 (2.3)	1.6 (2.1) 47 (55.3)	<0.001
Android:Gynoid Ratio, Mean (SD) Missing: n (%)	(0.14) 98 (63.2)	1.01 (0.15) 2 (2.3)	0.015 (0.085) 47 (55.3)	0.30
Trunk:Leg Ratio, Mean (SD) Missing: n (%)	0.95 (0.13) 98 (63.3)	0.97 (0.13) 2 (2.3)	0.016 (0.077) 47 (55.3)	0.20
Trunk:Limb Ratio, Mean (SD) Missing: n (%)	1.05 (0.20) 98 (63.3)	1.10 (0.23) 2 (2.3)	0.051 (0.11) 47 (55.3)	0.005
VAT Volume (in^3), Mean (SD) Missing: n (%)	592 (301) 98 (63.3)	633 (325) 2 (2.3)	88 (148) 47 (55.3)	<0.001

[1] Three additional CHS participants participated in the second visit without having completed the first. [2] p-values calculated using t-tests for continuous variables and McNemar–Bowker test for BMI category. [3] Fifty-seven participants completed the DEXA scan on a machine that provided additional body composition indices. Abbreviations: BMI, body mass index; FFMI, fat free mass index; VAT, visceral adipose tissue; SD, standard deviation.

3.1. Prediabetes/T2D

Positive change in HEI and DASH scores between the baseline and follow-up visits was associated with decreased risk for prediabetes/T2D at follow-up (Figure 1). A one-point increase in DASH score over the follow-up period was associated with a 64% (OR = 0.36, 95% CI: 0.17, 0.68) reduction in risk for prediabetes/T2D at follow-up, while a one-point increase in HEI between visits was associated with a 9% decrease in risk (OR = 0.91, 95% CI: 0.85, 0.96). When scaled by standard deviation of diet index, improvements in DASH diet score reduced the risk for prediabetes/T2D by a greater extent than the HEI (OR = 0.14, 95% CI: 0.03, 0.46; OR = 0.83, 95% CI: 0.72, 0.93, respectively). In the cross-sectional analysis of the follow-up visit, higher HEI and DASH scores were also associated with reduced risk for prediabetes/T2D. At baseline, only MDS was associated with reduced risk for prediabetes/T2D.

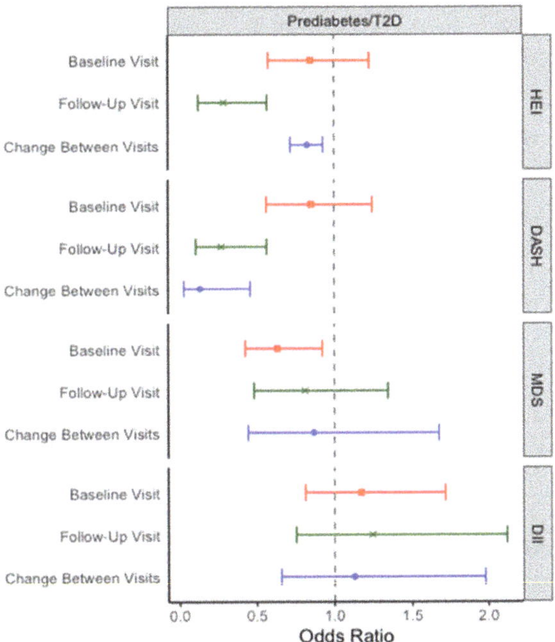

Figure 1. Coefficient plots for the effects of diet quality on prediabetes. "Baseline" and "follow-up" values are the result of cross-sectional analyses of diet quality score and risk of prediabetes/T2D at the same visit. The value for "change between visits" represents the risk of prediabetes/T2D at the follow-up visit associated with change in diet score between the baseline and the follow-up visit. Effects are standardized to one standard deviation of exposure. Covariates: *Baseline and follow-up models.* HEI, DASH, and DII models adjusted for age, sex, ethnicity, physical activity, and parental education. MDS models adjusted for energy intake, age, ethnicity, physical activity, and parental education. *Change between visits models.* Baseline and follow-up model covariates + baseline diet score. Abbreviations: DASH: Dietary Approaches to Stop Hypertension; DII: Dietary Inflammatory index; HEI: Healthy Eating Index—2015; MDS: Mediterranean Diet Score.

3.2. Fasting Glucose and Glucose Tolerance

There were no statistically significant cross-sectional associations between fasting glucose and any dietary index at either visit or between change in diet scores and change in fasting glucose between visits (Figure 2).

Higher HEI scores and higher MDS were associated with lower 2-h glucose values at baseline in the cross-sectional analyses (HEI: $\beta = -7.01$, 95% CI: -12.86, -1.16; MDS: $\beta = -7.43$, 95% CI: -13.25, -1.61) (Figure 2). Follow-up HEI and DASH scores were inversely associated with 2-h glucose at the same visit (HEI: $\beta = -8.64$, 95% CI: -16.16, -1.12; DASH: $\beta = -8.25$, 95% CI: -15.71, -0.78) and with glucose AUC (HEI: $\beta = -11.34$, 95% CI: -20.84, -1.84; DASH: $\beta = -10.99$, 95% CI: -20.44, -1.53).

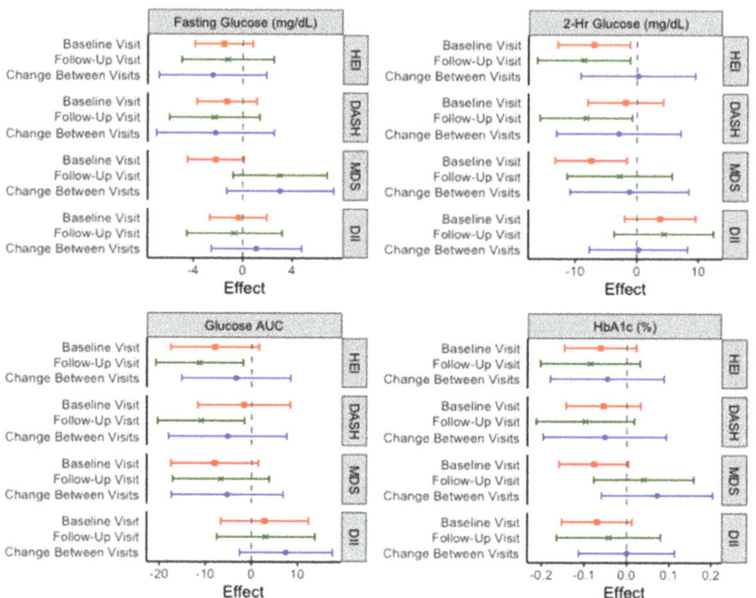

Figure 2. Coefficient plots for the effects of diet quality on glucose measurements. "Baseline" and "follow-up" values are the result of cross-sectional analyses of diet quality score and each outcome. The value for "change between visits" represents the association between the change in diet score between the baseline and the follow-up visit on the change in outcome between visits. Effects are scaled to one standard deviation of exposure. Covariates: *Baseline and follow-up models:* HEI, DASH, and DII models adjusted for age, sex, ethnicity, physical activity, and parental education. MDS models adjusted for energy intake, age, ethnicity, physical activity, and parental education. *Change between visits models:* Baseline and follow-up model covariates + baseline diet score. Abbreviations: DASH: Dietary Approaches to Stop Hypertension; DII: Dietary Inflammatory Index; HEI: Healthy Eating Index—2015; MDS: Mediterranean Diet Score.

3.3. Hemoglobin A1c

There were no statistically significant associations between HbA1c and any dietary index. However, there were consistent inverse relationships between higher HEI and DASH scores and HbA1c at both visits and between change in HEI or DASH and change in HbA1c between visits although these did not reach the threshold for statistical significance (Figure 2).

3.4. Body Composition

The DASH diet was consistently associated with several adiposity measures (Table 5). At the follow-up visit, higher DASH scores were associated with lower BMI ($\beta = -1.64$, 95% CI: -3.17, -0.11), body fat percent ($\beta = -1.79$, 95% CI: -3.01, -0.57), and fat mass to height ratio ($\beta = -1.09$, 95% CI: -3.27, -0.61) at the same visit, and increases in DASH between visits were also inversely associated with change in BMI ($\beta = -1.64$, 95% CI: -2.92, -0.36) and body fat percent ($\beta = -1.62$, 95% CI: -2.02, -0.17). Similar inverse associations were observed between DASH and measures of central adiposity, including trunk to limb ratio and VAT.

The DII was positively associated with body fat percent in the cross-sectional baseline analyses (Table 5). Though not statistically significant, the DII was also positively associated with most adiposity measurements at both visits, and positive change in DII was associated with positive changes in adiposity from baseline to follow-up.

Table 5. Estimated effect size and 95% CI for the effect of 1 standard deviation increase in diet score on body composition.

Diet	Outcome	Effect Estimate, β (95% CI)		
		Baseline [1]	Follow-Up [1]	Change between Visits [2]
Healthy Eating Index—2015 (HEI)				
	BMI (kg/m^2)	−0.62 (−1.45, 0.21)	−1.33 (−2.89, 0.24)	−0.38 (−1.62, 0.85)
	Body Fat (%)	−0.85 (−1.86, 0.16)	−1.09 (−2.37, 0.18)	0.40 (−0.92, 1.73)
	FFMI (kg/m^2)	−0.14 (−0.46, 0.17)	−0.46 (−1.04, 0.12)	−0.23 (−0.64, 0.18)
	Fat Mass:Height Ratio	−0.56 (−1.74, 0.62)	−0.73 (−1.68, 0.22)	−0.36 (−1.50, 0.78)
	Android:Gynoid Ratio	−0.045 (−0.087, −0.0036)	−0.043 (−0.071, −0.014)	−0.014 (−0.061, 0.034)
	Trunk:Leg Ratio	−0.040 (−0.077, −0.0028)	−0.035 (−0.060, −0.0087)	−0.0013 (−0.043, 0.041)
	Trunk:Limb Ratio	−0.052 (−0.11, 0.010)	−0.052 (−0.099, −0.0048)	−0.036 (−0.092, 0.020)
	VAT (in^3)	−65.78 (−161.45, 29.49)	−60.54 (−132.21, 11.13)	−48.05 (−123.33, 27.23)
Dietary Approaches to Stop Hypertension (DASH) Score				
	BMI (kg/m^2)	0.067 (−0.80, 0.94)	−1.64 (−3.17, −0.11)	−1.63 (−2.91, −0.35)
	Body Fat (%)	0.12 (−0.94, 1.18)	−1.79 (−3.01, −0.57)	−1.61 (−3.01, −0.21)
	FFMI (kg/m^2)	−0.036 (−0.36, 0.29)	−0.49 (−1.06, 0.088)	−0.41 (−0.85, 0.024)
	Fat Mass:Height Ratio	0.50 (−0.89, 1.88)	−1.09 (−2.02, −0.17)	−1.50 (−2.73, −0.27)
	Android:Gynoid Ratio	−0.015 (−0.066, 0.035)	−0.043 (−0.071, −0.015)	−0.047 (−0.098, 0.0045)
	Trunk:Leg Ratio	−0.023 (−0.068, 0.022)	−0.039 (−0.064, −0.014)	−0.037 (−0.084, 0.0097)
	Trunk:Limb Ratio	−0.018 (−0.093, 0.057)	−0.052 (−0.099, −0.0057)	−0.073 (−0.13, −0.011)
	VAT (in^3)	42.25 (−70.97, 155.46)	−76.57 (−146.46, −6.68)	−100.39 (−183.62, −17.17)
Mediterranean Diet Score (MDS)				
	BMI (kg/m^2)	−0.090 (−0.91, 0.73)	−0.71 (−2.28, 0.86)	0.27 (−0.95, 1.49)
	Body Fat (%)	−0.45 (−1.69, 0.79)	−0.48 (−2.35, 1.39)	1.24 (−0.062, 2.55)
	FFMI (kg/m^2)	0.078 (−0.32, 0.47)	0.075 (−0.57, 0.72)	−0.00040 (−0.42, 0.42)
	Fat Mass:Height Ratio	−0.37 (−1.49, 0.75)	−0.28 (−1.38, 0.83)	−0.081 (−1.11, 0.95)
	Android:Gynoid Ratio	0.00054 (−0.042, 0.043)	−0.0049 (−0.039, 0.030)	0.021 (−0.015, 0.057)
	Trunk:Leg Ratio	−0.030 (−0.065, 0.0037)	−0.0042 (−0.035, 0.027)	−0.0030 (−0.041, 0.035)
	Trunk:Limb Ratio	−0.044 (−0.10, 0.014)	−0.0073 (−0.062, 0.047)	−0.011 (−0.064, 0.042)
	VAT (in^3)	−21.86 (−109.41, 65.68)	−17.16 (−92.10, 57.79)	−25.82 (−98.45, 46.81)
Dietary Inflammatory Index (DII)				
	BMI (kg/m^2)	0.86 (0.044, 1.67)	−0.67 (−2.32, 0.97)	−0.21 (−1.24, 0.83)
	Body Fat (%)	2.04 (1.09, 2.99)	1.13 (−0.19, 2.45)	0.44 (−0.66, 1.54)
	FFMI (kg/m^2)	−0.073 (−0.38, 0.23)	−0.60 (−1.20, −0.0068)	−0.16 (−0.50, 0.18)
	Fat Mass:Height Ratio	0.88 (−0.23, 1.99)	−0.17 (−1.17, 0.84)	0.52 (−0.33, 1.37)
	Android:Gynoid Ratio	0.031 (−0.010, 0.072)	0.014 (−0.017, 0.045)	0.035 (0.0025, 0.068)
	Trunk:Leg Ratio	0.027 (−0.010, 0.063)	0.021 (−0.0070, 0.048)	0.017 (−0.014, 0.048)
	Trunk:Limb Ratio	0.028 (−0.033, 0.089)	0.023 (−0.027, 0.074)	0.029 (−0.014, 0.071)
	VAT (in^3)	47.00 (−44.96, 138.95)	−22.50 (−97.94, 52.94)	17.77 (−42.53, 78.08)

[1] Model A: outcome ~ diet score + covariates. [2] Model B: Δoutcome ~ Δdiet score + covariates. Model A covariates: HEI, DASH, and DII models adjusted for age, sex, ethnicity, physical activity, and parental education. MDS models adjusted for energy intake, age, ethnicity, physical activity, and parental education. Model B covariates: Model A covariates + baseline diet score. Effects were scaled to 1 standard deviation of exposure. Abbreviations: BMI: body mass index; FFMI: fat-free mass index; VAT: visceral adipose tissue.

3.5. Sensitivity Analyses

Results from the sensitivity analyses are reported in Supplemental Tables S1–S3. Models that did not adjust for physical activity had slightly larger effect estimates for the relationship between HEI and DASH and impaired glucose tolerance compared to models that did adjust for physical activity. There was little effect on risk for prediabetes/T2D, and the main findings were the same in the physical activity-adjusted and -unadjusted models. Adjustment for body fat percent also had little effect on the relationships between HEI or DASH and prediabetes/T2D, suggesting that it may not mediate the relationship between diet and prediabetes/T2D. However, in most cases, controlling for body fat percent attenuated the effects of each diet on all other glucose outcomes.

BMI, body fat percent, FFMI, fat mass to height ratio, and VAT were significantly associated with increased risk for prediabetes/T2D at all time points (Table S4). At the follow-up visit only, android to gynoid ratio, trunk to leg ratio, and trunk to limb ratio were also positively associated with prediabetes/T2D.

4. Discussion

We observed strong inverse associations both in cross-sectional and longitudinal analyses between the HEI and DASH diet and risk of prediabetes/T2D. We also found negative associations between the HEI and DASH diet and 2-h glucose, HbA1c, fasting glucose, and glucose AUC at both visits and in the longitudinal analysis though these relationships were not all statistically significant. The MDS was not consistently associated with prediabetes/T2D, glucose measurements, or body composition. We also observed inverse relationships between HEI, DASH, and MDS with measures of adiposity and body composition, suggesting that high diet quality may be protective against obesity and adverse accumulation of adipose tissue. The period between late adolescence to early adulthood is one of transition, where young people begin to live independently and gain more control of their lifestyles. However, there are limited assessments of change in diet quality during this transition [43], and these results emphasize the importance of considering diet quality in T2D risk within this age group.

To our knowledge, no other study has evaluated the longitudinal relationship between glucose dysregulation and HEI, DASH, MDS, and DII in young adults. Several meta-analyses have summarized the relationship between diet quality and type 2 diabetes, prediabetes, or other measures of glucose dysregulation in older adults. These analyses consistently report strong protective effects of healthy dietary patterns, including the DASH and HEI [10,13,15]. However, previous reviews found effects of similar magnitude between the HEI, DASH, and MDS [14], whereas we report a larger protective effect associated with increases in DASH diet adherence across both visits compared to either the HEI or MDS. The DII has been inconsistently associated with risk of T2D in older adults [17,18] though inflammation is involved in the pathogenesis of type 2 diabetes [44]. Like Vahid (2017), we observed positive associations between DII and impaired glucose intolerance and prediabetes.

Diet is also a risk factor for obesity, which is itself a significant driver of the T2D epidemic in both adults and youth [6,45,46], and increases in body fat greatly increase the risk for future diabetes [47]. Accumulation of visceral fat is also linked to T2D development and severity [48,49]. Our study found similar effects, with multiple adiposity indices significantly associated with increased risk of prediabetes/T2D. Our findings also suggest an inverse relationship between high diet quality and central obesity, with HEI and DASH consistently associated with android to gynoid fat ratio, trunk to limb fat ratios, and VAT. There also appeared to be positive associations between DII and adiposity and visceral fat measures. These findings suggest that high quality diets may reduce the risk of type 2 diabetes in part by reducing total body and visceral fat.

This study has several strengths. Participants were recruited from the Southern California Children's Health Study [30], which allowed detailed measures of glucose metabolism, diet, body composition, and lifestyle factors. OGTT and DEXA provide highly detailed information about glucose metabolism and body composition, respectively, beyond that of fasting plasma glucose, HbA1c, or BMI alone [50,51]. 2-h glucose and glucose AUC, for example, assess glucose tolerance, and impaired glucose tolerance is an early sign of glucose dysregulation and type 2 diabetes risk not often captured in clinical settings [52]. Additionally, exposures and outcomes were assessed at both visits, which allowed us to examine associations across time. Despite this, we note some limitations. Two systems were used to collect 24-h dietary recalls: the NDSR at baseline and the ASA24 at follow-up. We are not aware of any evidence that this difference would introduce bias away from the null, and any misclassification of diet is expected to be nondifferential and independent of prediabetes/T2D status. It is also common for studies involving multiple cohorts to integrate different diet assessment measures [53,54]. There is a possibility that residual confounding contributed to our reported effects; family history of T2D, maternal obesity, and low birthweight are also associated with young-onset T2D though they are less likely to be associated with diet. However, the magnitude of the relationships we report are large, and any confounding by these or other factors are unlikely to account for the entire effect. Additionally, our sample size for the longitudinal analysis was 85, limiting the statistical power to detect significant relationships. Limitations of one of the DEXA machines used at baseline also limited the available sample size for some adiposity measurements (e.g., android to gynoid fat ratio, trunk to limb fat ratio). However, power was sufficient to identify strong, statistically significant, protective effects of high-quality diets on prediabetes risk.

The COVID-19 pandemic may also have affected our recruitment efforts for the follow-up visit. Our recruitment began as the SARS-CoV-2 virus (COVID-19) was declared first a Public Health Emergency and then a pandemic [55]. The resulting disruptions to daily life would have affected our participants and likely impacted lifestyle factors such as physical activity, sleep, and eating habits as well as stress, social supports, and physical health, all of which may affect non-communicable disease risk [55–57]. It is possible that the observed decreases in diet quality between the baseline and follow-up visits may be, in part, due to the pandemic. Even if some of the change in diet were due to changes in lifestyle associated with the COVID-19 pandemic, our findings emphasize the importance of maintaining a healthy diet to reduce the risk for T2D.

Our results indicate that improvements in adherence to the HEI and DASH dietary patterns may reduce risk for T2D. Though both measure diet quality, the construction of each index emphasizes different nutrients and food groups, and there are several ways in which an individual may improve their score and overall diet quality. For example, the HEI rewards greater adherence to the USDA Dietary Guidelines for Americans with higher scores on a 100-point scale [36]. To improve a HEI score, one has several options: (1) increase intake of one or several food groups (fruit, vegetables, seafood, etc.) to the levels recommend by the USDA; (2) reduce intake of added sugars and salt as recommended by the USDA; or (3) reduce the proportion of total grains that come from refined sources or increase the proportion of dietary fats that are mono- or polyunsaturated [58]. Similarly, improvements in DASH diet score could be achieved by reducing consumption of saturated fat, cholesterol, or sodium, or by increasing fiber, magnesium, potassium, and calcium intake [37]. By encouraging changes to overall dietary patterns rather than emphasizing specific foods or nutrients (i.e., kilocalories, sugar-sweetened beverages), individuals may have more flexibility in their choice of dietary habits to alter or methods of alteration, leading to more successful behavior change [59,60].

5. Conclusions

Late adolescence to early adulthood is a period of significant change and represents an important window in which to establish lifelong habits [29]. To our knowledge, this study is one of few to evaluate the impact of dietary changes on glucose regulation in people between the ages of 18 and 30. We found that adherence to the DASH diet and USDA Dietary Guidelines is associated with reduced risk for prediabetes and better glucose tolerance. Improvement in DASH or HEI scores over the follow-up period was also associated with lower risk for prediabetes or type 2 diabetes, with the strongest effects observed for the DASH diet. These findings indicate that the DASH dietary pattern may be a promising target for diabetes prevention efforts in young adults.

Supplementary Materials: The following supporting information can be downloaded at: https://www.mdpi.com/article/10.3390/nu14183734/s1, Figure S1: Flowchart for study recruitment; Table S1: Results (effects and 95% CIs) for sensitivity analyses at the baseline visit; Table S2: Results (effects and 95% CIs) for sensitivity analyses at the follow-up visit; Table S3: Results (effects and 95% CIs) for sensitivity analyses for the effects of change in diet score between visits; Table S4: Estimated effect size and 95% CI for the relationship between body composition and risk for prediabetes/type 2 diabetes).

Author Contributions: Conceptualization, E.C., J.G. and L.C.; data curation, E.C., J.G., W.B.P. and S.R.; formal analysis, E.C., W.B.P. and Y.L.; funding acquisition, L.C.; investigation, E.C., J.G. and S.R.; methodology, E.C., J.G., W.B.P., Z.C., T.L.A., D.V.C. and L.C.; resources, F.G., Z.C., T.L.A. and L.C.; supervision, J.G., B.B., F.G., M.I.G., Z.C., T.L.A., D.V.C. and L.C.; visualization, E.C. and J.G.; writing—original draft, E.C.; writing—review and editing, J.G., W.B.P., S.R., Y.L., B.B., F.G., M.I.G., Z.C., T.L.A., D.V.C. and L.C. All authors have read and agreed to the published version of the manuscript.

Funding: The results reported here in correspond to specific aims of grand R01ES029944 from the National Institute of Environmental Health Science (NIEHS). Funding for the MetaAir study came from the Southern California Children's Environmental Health Center grants funded by NIEHS (5P01ES022845-03, P30ES007048, 5P01ES011627), the United States Environmental Protection Agency (RD83544101), and the Hastings Foundation. Additional funding from NIEHS supported E.C. and J.G. (T32ES013678), Z.C. (R00ES027870, R00ES027853), T.L.A. (R00ES027853), D.V.C. (R01ES030691, R21ES029681, R01ES030364, R01ES016813), and L.C. (R01ES030691, R21ES029681, R01ES030364, R21ES028903). Other sources of funding included the National Institute for Diabetes and Digestive and Kidney Diseases (F31DK134198, W.B.P.; P30DK048522, D.V.C.), the National Cancer Institute (P01CA196569 and R01CA140561, D.V.C.), and the USC Center for Translational Research on Environmental Health (S.R.).

Institutional Review Board Statement: The study was conducted in accordance with the Declaration of Helsinki and approved by the Institutional Review Board of the University of Southern California (HS-19-00338; 3 June 2019).

Informed Consent Statement: Informed consent was obtained from all subjects involved in the study.

Data Availability Statement: The data presented in this study are available on request from the corresponding author. The data are not publicly available to protect participants' identifiable information.

Conflicts of Interest: The authors declare no conflict of interest. The funders had no role in the design of the study; in the collection, analyses, or interpretation of data; in the writing of the manuscript; or in the decision to publish the results.

References

1. Tabák, A.G.; Herder, C.; Rathmann, W.; Brunner, E.J.; Kivimäki, M. Prediabetes: A high-risk state for diabetes development. *Lancet* **2012**, *379*, 2279–2290. [CrossRef]
2. Andes, L.J.; Cheng, Y.J.; Rolka, D.B.; Gregg, E.W.; Imperatore, G. Prevalence of Prediabetes Among Adolescents and Young Adults in the United States, 2005–2016. *JAMA Pediatr.* **2020**, *174*, e194498. [CrossRef] [PubMed]
3. Liu, J.; Li, Y.; Zhang, D.; Yi, S.S.; Liu, J. Trends in Prediabetes Among Youths in the US From 1999 Through 2018. *JAMA Pediatr.* **2022**, *176*, 608–611. [CrossRef] [PubMed]
4. Morris, D.H.; Khunti, K.; Achana, F.; Srinivasan, B.; Gray, L.J.; Davies, M.J.; Webb, D. Progression rates from HbA1c 6.0–6.4% and other prediabetes definitions to type 2 diabetes: A meta-analysis. *Diabetologia* **2013**, *56*, 1489–1493. [CrossRef]
5. Divers, J.; Mayer-Davis, E.J.; Lawrence, J.M.; Isom, S.; Dabelea, D.; Dolan, L.; Imperatore, G.; Marcovina, S.; Pettitt, D.J.; Pihoker, C.; et al. Trends in Incidence of Type 1 and Type 2 Diabetes Among Youths-Selected Counties and Indian Reservations, United States, 2002–2015. *MMWR Morb. Mortal. Wkly. Rep.* **2020**, *69*, 161–165. [CrossRef]
6. Shah, A.S.; Nadeau, K.J. The changing face of paediatric diabetes. *Diabetologia* **2020**, *63*, 683–691. [CrossRef]
7. Magliano, D.J.; Sacre, J.W.; Harding, J.L.; Gregg, E.W.; Zimmet, P.Z.; Shaw, J.E. Young-onset type 2 diabetes mellitus-implications for morbidity and mortality. *Nat. Rev. Endocrinol.* **2020**, *16*, 321–331. [CrossRef]

8. Lascar, N.; Brown, J.; Pattison, H.; Barnett, A.H.; Bailey, C.J.; Bellary, S. Type 2 diabetes in adolescents and young adults. *Lancet Diabetes Endocrinol.* **2018**, *6*, 69–80. [CrossRef]
9. American Diabetes Association. 3. Prevention or Delay of Type 2 Diabetes: Standards of Medical Care in Diabetes—2020. *Diabetes Care* **2020**, *43* (Suppl. 1), S32–S36. [CrossRef]
10. Neuenschwander, M.; Ballon, A.; Weber, K.S.; Norat, T.; Aune, D.; Schwingshackl, L.; Schlesinger, S. Role of diet in type 2 diabetes incidence: Umbrella review of meta-analyses of prospective observational studies. *BMJ* **2019**, *366*, l2368. [CrossRef]
11. Schwingshackl, L.; Hoffmann, G.; Lampousi, A.M.; Knüppel, S.; Iqbal, K.; Schwedhelm, C.; Bechthold, A.; Schlesinger, S.; Boeing, H. Food groups and risk of type 2 diabetes mellitus: A systematic review and meta-analysis of prospective studies. *Eur. J. Epidemiol.* **2017**, *32*, 363–375. [CrossRef]
12. Toi, P.L.; Anothaisintawee, T.; Chaikledkaew, U.; Briones, J.R.; Reutrakul, S.; Thakkinstian, A. Preventive Role of Diet Interventions and Dietary Factors in Type 2 Diabetes Mellitus: An Umbrella Review. *Nutrients* **2020**, *12*, 2722. [CrossRef]
13. De Koning, L.; Chiuve, S.E.; Fung, T.T.; Willett, W.C.; Rimm, E.B.; Hu, F.B. Diet-quality scores and the risk of type 2 diabetes in men. *Diabetes Care* **2011**, *34*, 1150–1156. [CrossRef]
14. Jannasch, F.; Kröger, J.; Schulze, M.B. Dietary Patterns and Type 2 Diabetes: A Systematic Literature Review and Meta-Analysis of Prospective Studies. *J. Nutr.* **2017**, *147*, 1174–1182. [CrossRef]
15. Qian, F.; Liu, G.; Hu, F.B.; Bhupathiraju, S.N.; Sun, Q. Association Between Plant-Based Dietary Patterns and Risk of Type 2 Diabetes: A Systematic Review and Meta-analysis. *JAMA Intern. Med.* **2019**, *179*, 1335–1344. [CrossRef]
16. Salas-Salvadó, J.; Bulló, M.; Estruch, R.; Ros, E.; Covas, M.I.; Ibarrola-Jurado, N.; Corella, D.; Arós, F.; Gómez-Gracia, E.; Ruiz-Gutiérrez, V.; et al. Prevention of diabetes with Mediterranean diets: A subgroup analysis of a randomized trial. *Ann. Intern. Med.* **2014**, *160*, 1–10. [CrossRef]
17. Vahid, F.; Shivappa, N.; Karamati, M.; Naeini, A.J.; Hebert, J.R.; Davoodi, S.H. Association between Dietary Inflammatory Index (DII) and risk of prediabetes: A case-control study. *Appl. Physol. Nutr. Metab.* **2017**, *42*, 399–404. [CrossRef]
18. Laouali, N.; Mancini, F.R.; Hajji-Louati, M.; El Fatouhi, D.; Balkau, B.; Boutron-Ruault, M.-C.; Bonnet, F.; Fagherazzi, G. Dietary inflammatory index and type 2 diabetes risk in a prospective cohort of 70,991 women followed for 20 years: The mediating role of BMI. *Diabetologia* **2019**, *62*, 2222–2232. [CrossRef]
19. Esposito, K.; Chiodini, P.; Maiorino, M.I.; Bellastella, G.; Panagiotakos, D.; Giugliano, D. Which diet for prevention of type 2 diabetes? A meta-analysis of prospective studies. *Endocrine* **2014**, *47*, 107–116. [CrossRef]
20. Alhazmi, A.; Stojanovski, E.; McEvoy, M.; Garg, M.L. The association between dietary patterns and type 2 diabetes: A systematic review and meta-analysis of cohort studies. *J. Hum. Nutr. Diet.* **2014**, *27*, 251–260. [CrossRef]
21. Barnes, T.L.; Crandell, J.L.; Bell, R.A.; Mayer-Davis, E.J.; Dabelea, D.; Liese, A.D.; for the SEARCH for Diabetes in Youth Study Group. Change in DASH diet score and cardiovascular risk factors in youth with type 1 and type 2 diabetes mellitus: The SEARCH for Diabetes in Youth Study. *Nutr. Diabetes* **2013**, *3*, e91. [CrossRef] [PubMed]
22. Gow, M.L.; Garnett, S.P.; Baur, L.A.; Lister, N.B. The Effectiveness of Different Diet Strategies to Reduce Type 2 Diabetes Risk in Youth. *Nutrients* **2016**, *8*, 486. [CrossRef] [PubMed]
23. Kim, R.J.; Lopez, R.; Snair, M.; Tang, A. Mediterranean diet adherence and metabolic syndrome in US adolescents. *Int. J. Food Sci. Nutr.* **2021**, *72*, 537–547. [CrossRef] [PubMed]
24. Cañete, R.; Gil-Campos, M.; Aguilera, C.M.; Gil, A. Development of insulin resistance and its relation to diet in the obese child. *Eur. J. Nutr.* **2007**, *46*, 181–187. [CrossRef] [PubMed]
25. Hirahatake, K.M.; Jacobs, D.R., Jr.; Shikany, J.M.; Jiang, L.; Wong, N.D.; Odegaard, A.O. Cumulative average dietary pattern scores in young adulthood and risk of incident type 2 diabetes: The CARDIA study. *Diabetologia* **2019**, *62*, 2233–2244. [CrossRef] [PubMed]
26. Cha, E.; Pasquel, F.J.; Yan, F.; Jacobs, D.R., Jr.; Dunbar, S.B.; Umpierrez, G.; Choi, Y.; Shikany, J.M.; Bancks, M.P.; Reis, J.P.; et al. Characteristics associated with early- vs. later-onset adult diabetes: The CARDIA study. *Diabetes Res. Clin. Pract.* **2021**, *182*, 109144. [CrossRef] [PubMed]
27. Bantle, A.E.; Chow, L.S.; Steffen, L.M.; Wang, Q.; Hughes, J.; Durant, N.H.; Ingram, K.H.; Reis, J.P.; Schreiner, P.J. Association of Mediterranean diet and cardiorespiratory fitness with the development of pre-diabetes and diabetes: The Coronary Artery Risk Development in Young Adults (CARDIA) study. *BMJ Open Diabetes Res. Care* **2016**, *4*, e000229. [CrossRef]
28. Choi, Y.; Larson, N.; Gallaher, D.D.; Odegaard, A.O.; Rana, J.S.; Shikany, J.M.; Steffen, L.M.; Jacobs, D.R., Jr. A Shift Toward a Plant-Centered Diet From Young to Middle Adulthood and Subsequent Risk of Type 2 Diabetes and Weight Gain: The Coronary Artery Risk Development in Young Adults (CARDIA) Study. *Diabetes Care* **2020**, *43*, 2796–2803. [CrossRef]
29. Nelson, M.C.; Story, M.; Larson, N.I.; Neumark-Sztainer, D.; Lytle, L.A. Emerging Adulthood and College-aged Youth: An Overlooked Age for Weight-related Behavior Change. *Obesity* **2008**, *16*, 2205–2211. [CrossRef]
30. McConnell, R.; Shen, E.; Gilliland Frank, D.; Jerrett, M.; Wolch, J.; Chang, C.-C.; Lurmann, F.; Berhane, K. A Longitudinal Cohort Study of Body Mass Index and Childhood Exposure to Secondhand Tobacco Smoke and Air Pollution: The Southern California Children's Health Study. *Environ. Health Perspect.* **2015**, *123*, 360–366. [CrossRef]
31. Kim, J.S.; Chen, Z.; Alderete, T.L.; Toledo-Corral, C.; Lurmann, F.; Berhane, K.; Gilliland, F.D. Associations of air pollution, obesity and cardiometabolic health in young adults: The Meta-AIR study. *Environ. Int.* **2019**, *133*, 105180. [CrossRef]
32. Floch, J.-P.L.; Escuyer, P.; Baudin, E.; Baudon, D.; Perlemuter, L. Blood Glucose Area Under the Curve: Methodological Aspects. *Diabetes Care* **1990**, *13*, 172–175. [CrossRef]

33. American Diabetes Association. 2. Classification and Diagnosis of Diabetes: Standards of Medical Care in Diabetes—2021. *Diabetes Care* **2020**, *44* (Suppl. 1), S15–S33. [CrossRef]
34. Hoffmann, K.; Boeing, H.; Dufour, A.; Volatier, J.L.; Telman, J.; Virtanen, M.; Becker, W.; De Henauw, S.; for the EFCOSUM Group. Estimating the distribution of usual dietary intake by short-term measurements. *Eur. J. Clin. Nutr.* **2002**, *56*, S53–S62. [CrossRef]
35. Park, Y.; Dodd, K.W.; Kipnis, V.; Thompson, F.E.; Potischman, N.; Schoeller, D.A.; Baer, D.J.; Midthune, D.; Troiano, R.P.; Bowles, H.; et al. Comparison of self-reported dietary intakes from the Automated Self-Administered 24-h recall, 4-d food records, and food-frequency questionnaires against recovery biomarkers. *Am. J. Clin. Nutr.* **2018**, *107*, 80–93. [CrossRef]
36. Krebs-Smith, S.M.; Pannucci, T.E.; Subar, A.F.; Kirkpatrick, S.I.; Lerman, J.L.; Tooze, J.A.; Wilson, M.M.; Reedy, J. Update of the Healthy Eating Index: HEI-2015. *J. Acad. Nutr. Diet.* **2018**, *118*, 1591–1602. [CrossRef]
37. Mellen, P.B.; Gao, S.K.; Vitolins, M.Z.; Goff, D.C., Jr. Deteriorating Dietary Habits Among Adults with Hypertension: DASH Dietary Accordance, NHANES 1988–1994 and 1999–2004. *Arch. Intern. Med.* **2008**, *168*, 308–314. [CrossRef]
38. Trichopoulou, A.; Costacou, T.; Bamia, C.; Trichopoulos, D. Adherence to a Mediterranean diet and survival in a Greek population. *N. Engl. J. Med.* **2003**, *348*, 2599–2608. [CrossRef]
39. Shivappa, N.; Steck, S.E.; Hurley, T.G.; Hussey, J.R.; Hébert, J.R. Designing and developing a literature-derived, population-based dietary inflammatory index. *Public Health Nutr.* **2014**, *17*, 1689–1696. [CrossRef]
40. Food and Drug Administration. Final determination regarding partially hydrogenated oils. *Fed. Regist.* **2015**, *80*, 34650–34670.
41. Craig, C.L.; Marshall, A.L.; Sjöström, M.; Bauman, A.E.; Booth, M.L.; Ainsworth, B.E.; Pratt, M.; Ekelund, U.; Yngve, A.; Sallis, J.F.; et al. International Physical Activity Questionnaire: 12-Country Reliability and Validity. *Med. Sci. Sports Exerc.* **2003**, *35*, 1381–1395. [CrossRef]
42. Schisterman, E.F.; Cole, S.R.; Platt, R.W. Overadjustment bias and unnecessary adjustment in epidemiologic studies. *Epidemiology* **2009**, *20*, 488–495. [CrossRef]
43. Winpenny, E.M.; Penney, T.L.; Corder, K.; White, M.; van Sluijs, E.M.F. Change in diet in the period from adolescence to early adulthood: A systematic scoping review of longitudinal studies. *Int. J. Behav. Nutr. Phys. Act.* **2017**, *14*, 60. [CrossRef]
44. Esser, N.; Legrand-Poels, S.; Piette, J.; Scheen, A.J.; Paquot, N. Inflammation as a link between obesity, metabolic syndrome and type 2 diabetes. *Diabetes Res. Clin. Pract.* **2014**, *105*, 141–150. [CrossRef]
45. Pulgaron, E.R.; Delamater, A.M. Obesity and type 2 diabetes in children: Epidemiology and treatment. *Curr. Diab. Rep.* **2014**, *14*, 508. [CrossRef]
46. Maggio, C.A.; Pi-Sunyer, F.X. Obesity and type 2 diabetes. *Endocrinol. Metab. Clin. N. Am.* **2003**, *32*, 805–822. [CrossRef]
47. Zhao, T.; Lin, Z.; Zhu, H.; Wang, C.; Jia, W. Impact of body fat percentage change on future diabetes in subjects with normal glucose tolerance. *IUBMB Life* **2017**, *69*, 947–955. [CrossRef]
48. Chen, Y.; He, D.; Yang, T.; Zhou, H.; Xiang, S.; Shen, L.; Wen, J.; Chen, S.; Peng, S.; Gan, Y. Relationship between body composition indicators and risk of type 2 diabetes mellitus in Chinese adults. *BMC Public Health* **2020**, *20*, 452. [CrossRef] [PubMed]
49. Gupta, P.; Lanca, C.; Gan, A.T.L.; Soh, P.; Thakur, S.; Tao, Y.; Kumari, N.; Man, R.E.K.; Fenwick, E.K.; Lamoureux, E.L. The Association between Body Composition using Dual energy X-ray Absorptiometry and Type-2 Diabetes: A Systematic Review and Meta-Analysis of Observational studies. *Sci. Rep.* **2019**, *9*, 12634. [CrossRef] [PubMed]
50. Jesudason, D.R.; Dunstan, K.; Leong, D.; Wittert, G.A. Macrovascular Risk and Diagnostic Criteria for Type 2 Diabetes: Implications for the use of FPG and HbA1c for cost-effective screening. *Diabetes Care* **2003**, *26*, 485–490. [CrossRef] [PubMed]
51. Messina, C.; Albano, D.; Gitto, S.; Tofanelli, L.; Bazzocchi, A.; Ulivieri, F.M.; Guglielmi, G.; Sconfienza, L.M. Body composition with dual energy X-ray absorptiometry: From basics to new tools. *Quant. Imaging Med. Surg.* **2020**, *10*, 1687–1698. [CrossRef]
52. Bartoli, E.; Fra, G.P.; Carnevale Schianca, G.P. The oral glucose tolerance test (OGTT) revisited. *Eur. J. Intern. Med.* **2011**, *22*, 8–12. [CrossRef]
53. Maitre, L.; de Bont, J.; Casas, M.; Robinson, O.; Aasvang, G.M.; Agier, L.; Andrušaitytė, S.; Ballester, F.; Basagaña, X.; Borràs, E.; et al. Human Early Life Exposome (HELIX) study: A European population-based exposome cohort. *BMJ Open* **2018**, *8*, e021311. [CrossRef]
54. Smith-Warner, S.A.; Spiegelman, D.; Ritz, J.; Albanes, D.; Beeson, W.L.; Bernstein, L.; Berrino, F.; van den Brandt, P.A.; Buring, J.E.; Cho, E.; et al. Methods for pooling results of epidemiologic studies: The Pooling Project of Prospective Studies of Diet and Cancer. *Am. J. Epidemiol.* **2006**, *163*, 1053–1064. [CrossRef]
55. Machhi, J.; Herskovitz, J.; Senan, A.M.; Dutta, D.; Nath, B.; Oleynikov, M.D.; Blomberg, W.R.; Meigs, D.D.; Hasan, M.; Patel, M.; et al. The Natural History, Pathobiology, and Clinical Manifestations of SARS-CoV-2 Infections. *J. Neuroimmune Pharmacol.* **2020**, *15*, 359–386. [CrossRef]
56. Karatzi, K.; Poulia, K.A.; Papakonstantinou, E.; Zampelas, A. The Impact of Nutritional and Lifestyle Changes on Body Weight, Body Composition and Cardiometabolic Risk Factors in Children and Adolescents during the Pandemic of COVID-19: A Systematic Review. *Children* **2021**, *8*, 1130. [CrossRef]
57. Kreutz, R.; Dobrowolski, P.; Prejbisz, A.; Algharably, E.A.E.; Bilo, G.; Creutzig, F.; Grassi, G.; Kotsis, V.; Lovic, D.; Lurbe, E.; et al. Lifestyle, psychological, socioeconomic and environmental factors and their impact on hypertension during the coronavirus disease 2019 pandemic. *J. Hypertens.* **2021**, *39*, 1077–1089. [CrossRef]
58. Millen, B.E.; Abrams, S.; Adams-Campbell, L.; Anderson, C.A.; Brenna, J.T.; Campbell, W.W.; Clinton, S.; Hu, F.; Nelson, M.; Neuhouser, M.L.; et al. The 2015 Dietary Guidelines Advisory Committee Scientific Report: Development and Major Conclusions. *Adv. Nutr.* **2016**, *7*, 438–444. [CrossRef]

59. Gardner, C.D.; Trepanowski, J.F.; Del Gobbo, L.C.; Hauser, M.E.; Rigdon, J.; Ioannidis, J.P.A.; Desai, M.; King, A.C. Effect of Low-Fat vs Low-Carbohydrate Diet on 12-Month Weight Loss in Overweight Adults and the Association With Genotype Pattern or Insulin Secretion: The DIETFITS Randomized Clinical Trial. *JAMA* **2018**, *319*, 667–679. [CrossRef]
60. Koliaki, C.; Spinos, T.; Spinou, M.; Brinia, M.-E.; Mitsopoulou, D.; Katsilambros, N. Defining the Optimal Dietary Approach for Safe, Effective and Sustainable Weight Loss in Overweight and Obese Adults. *Healthcare* **2018**, *6*, 73. [CrossRef]

Article

The Relationship between Dietary Patterns and High Blood Glucose among Adults Based on Structural Equation Modelling

Yuanyuan Wang [1,†], Wei Xie [2,†], Ting Tian [2], Jingxian Zhang [2], Qianrang Zhu [2], Da Pan [1], Dengfeng Xu [1], Yifei Lu [1], Guiju Sun [1] and Yue Dai [1,2,*]

1 Key Laboratory of Environmental Medicine and Engineering of Ministry of Education, Department of Nutrition and Food Hygiene, School of Public Health, Southeast University, Nanjing 210009, China
2 Institute of Food Safety and Assessment, Jiangsu Provincial Center for Disease Control and Prevention, Nanjing 210009, China
* Correspondence: 18915999341@163.com; Tel./Fax: +86-25-83759341
† These authors contributed equally to this work.

Abstract: Abstract: AimsThe aim of this study was to examine the association between dietary patterns and high blood glucose in Jiangsu province of China by using structural equation modelling (SEqM). **Methods:** Participants in this cross-sectional study were recruited through the 2015 Chinese Adult Chronic Disease and Nutrition Surveillance Program in Jiangsu province using a multistage stratified cluster random sampling method. Dietary patterns were defined by exploratory factor analysis (EFA). Confirmatory factor analysis (CFA) was used to test the fitness of EFA. SEqM was used to investigate the association between dietary patterns and high blood glucose. **Results:** After exclusion, 3137 participants with complete information were analysed for this study. The prevalence of high blood glucose was 9.3% and 8.1% in males and females, respectively. Two dietary patterns: the modern dietary pattern (i.e., high in red meats and its products, vegetables, seafood, condiments, fungi and algae, main grains and poultry; low in other grains, tubers and preserves), and the fruit–milk dietary pattern (i.e., high in milk and its products, fruits, eggs, nuts and seeds and pastry snacks, but low in vegetable oils) were established. Modern dietary pattern was found to be positively associated with high blood glucose in adults in Jiangsu province (multivariate logistic regression: OR = 1.561, 95% CI: 1.025~2.379; SEqM: $\beta = 0.127$, $p < 0.05$). **Conclusion:** The modern dietary pattern—high intake of red meats—was significantly associated with high blood glucose among adults in Jiangsu province of China, while the fruit–milk dietary pattern was not significantly associated with high blood glucose.

Keywords: dietary pattern; high blood glucose; structural equation modelling

Citation: Wang, Y.; Xie, W.; Tian, T.; Zhang, J.; Zhu, Q.; Pan, D.; Xu, D.; Lu, Y.; Sun, G.; Dai, Y. The Relationship between Dietary Patterns and High Blood Glucose among Adults Based on Structural Equation Modelling. *Nutrients* **2022**, *14*, 4111. https://doi.org/10.3390/nu14194111

Academic Editor: Giuseppe Della Pepa

Received: 6 September 2022
Accepted: 29 September 2022
Published: 3 October 2022

Publisher's Note: MDPI stays neutral with regard to jurisdictional claims in published maps and institutional affiliations.

Copyright: © 2022 by the authors. Licensee MDPI, Basel, Switzerland. This article is an open access article distributed under the terms and conditions of the Creative Commons Attribution (CC BY) license (https://creativecommons.org/licenses/by/4.0/).

1. Introduction

Diabetes mellitus (DM) is a chronic metabolic disease characterised by elevated blood glucose levels. As the disease progresses, it can further damage the heart, eyes and kidneys [1–4]. There are approximately 536.6 million people living with diabetes worldwide, and 6.7 million people die from it each year [5]. In the United States, approximately 32.2 million adults have diabetes, and 36.3 million are expected to have diabetes in 2045 [5]. In China, diabetes is considered a major health issue, with prevalence significantly increasing among adults 18 years and older, rising from 9.7% in 2012 to 11.9% in 2018 [6]. Diabetes is susceptible to unhealthy lifestyles, such as smoking, alcohol consumption and unhealthy eating habits [7,8].

Over the past few decades, studies have shown that diet acted as a major factor in the development of DM. Epidemiological studies suggested that approximately 80% of DM can be prevented through healthy dietary habits such as regular consumption of fruits and vegetables and reduced intake of saturated fat, sodium and sugar-sweetened drinks [9,10]. In a meta-analysis, vitamin D supplementation reduced the risk of type 2 diabetes (T2DM) and increased the rate of return to normal blood glucose in individuals with prodromal DM [11]. However, because of the dietary complexity of different populations and the potential of food–food or food–component interactions, it could be difficult to evaluate the effect of a single or a few foods or nutrients on DM [12,13]. One study has shown that the western dietary pattern, as determined by the Gaussian graphical models, had a positive association with the risk of T2DM in women [14]. In addition, the Mediterranean diet may reduce the risk of cardiovascular disease (CVD) in patients with DM [15].

A variety of studies on analysing dietary pattern methods have emerged in recent years [13,16]. Among them, structural equation modelling (SEqM) is an appropriate approach to statistics that merges the methods of factor analysis and path analysis to determine the direct and indirect correlations between potential and observed variables. It can take both errors and individual differences into account [17,18].

To our knowledge, there are no studies explaining the association of direct and indirect associations with high blood glucose as well as socio-demographics in Jiangsu province, China. Therefore, the aims of this study were as follows: (i) to determine the final dietary pattern by exploratory factor analysis (EFA) and confirmatory factor analysis (CFA), and (ii) to examine the effect of dietary patterns on high blood glucose among adults in Jiangsu province, China.

2. Participants and Methods

2.1. Study Population

The China Adult Chronic Disease and Nutrition Surveillance Project (2015) in Jiangsu province covered thirteen surveillance sites, including Qinhuai, Chongan, Yunlong, Wujin, Wuzhong, Zhangjiagang, Rugao, Donghai, Jinhu, Xiangshui, Hanjiang, Jingkou and Jiangyan. Recruitment of participants using multistage stratified cluster random sampling methods: ① at each surveillance site, three streets/towns were randomly selected by using a cluster sampling method proportional to the population size; ② two further villages/communities were randomly selected in each street/township using a cluster sampling method; ③ in each village/neighbourhood, one village group was selected by using a simple random sampling method (at least 60 households); ④ 45 households were randomly selected, and all residents in the households were enrolled in the survey. After screening, as shown in Figure 1, a total of 3137 participants (54.8% female, n = 1718) aged ≥18 years old with complete 3-day and 24 h-dietary recall data were included in this study.

2.2. Anthropometric Measurement

All participants were asked to wear light clothing and no shoes while taking anthropometric measurements. The TANITA HD-390 electronic weight scale (Dongwan, China) was used for weight measurement. Height measurement was performed using a TZG sit height gauge (Wuxi, China). All measurements were carried out twice to ensure the stability of the measurement results [16]. The body mass index (BMI) standard for the Chinese population was used to define the BMI classification for this study [17]. Central obesity was defined as ≥90 cm waist circumference in men and ≥85 cm in women.

2.3. Biochemical Indicator

Participants' blood was collected early in the morning (8–12 h fasting). Oral confirmation of fasting was obtained from the participant prior to the blood sample collection. Then, the blood was centrifuged and fasting blood glucose (FBG) was measured using the glucose oxidase method. High blood glucose was defined as: (1) self-reported diabetes or diabetes that has been diagnosed and treated by doctors (including herbs, western medicine and insulin injections); (2) FBG ≥ 7.0 mmol/L.

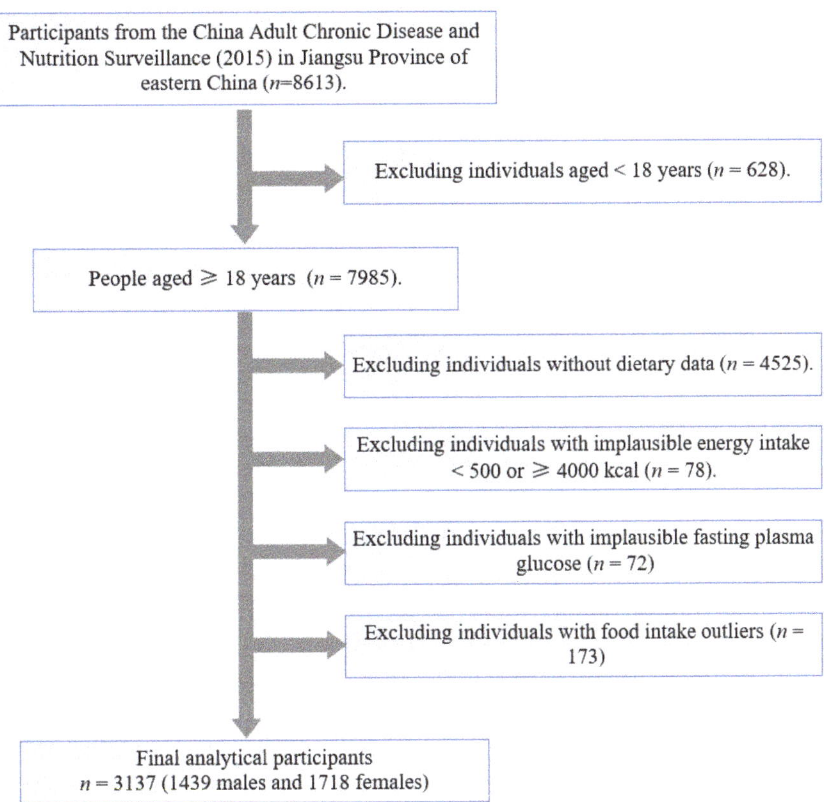

Figure 1. Flow chart of participants included in the study.

2.4. Dietary Assessment

We used a 3-day intake of various consumed foods, including alcohol consumption, various condiments, etc., to assess the actual daily intake of individuals. The 3-day and 24-h dietary recall and food weighing method were the keys to obtaining the individual dietary intake. Professionally trained investigators recorded the foods participants had eaten in the past 24 h during the first home visit and taught them how to record their food intake. All food data were obtained through face-to-face interviews. Participants were required to record their food consumption for 3 consecutive days (including 2 weekdays and 1 weekend day). During the survey, food models and household measurement tools were used to help participants estimate their portion sizes. The collected foods were then combined into food groups according to the Chinese food composition table (2002). Using the Chinese Dietary Guidelines and combined with the dietary characteristics of the Jiangsu population, we classified the food into 22 food groups, as shown in Supplementary Table S1.

2.5. Dietary Pattern Analysis

For dietary pattern analysis, we mainly used factor analysis for exploration. Factor analysis includes two parts: confirmatory factor analysis and exploratory factor analysis. In this study, EFA was first used to summarise the main dietary structure of the study population. The Kaiser–Meyer–Olkin index (KMO) and Bartlett's spherical test suggested that the data were suited for EFA. Shared factors were extracted using principal component analysis, and their number was determined based on the eigenvalues >1.3, scree plot and interpretability of derived factors. The maximum variance orthogonal rotation method was used at the same time to make each common factor more obvious and professional.

The absolute value of the factor loadings >0.25 was used to determine the main food composition of each common factor. After that, we put the major foods with absolute values of factor loadings >0.25 into the validation factor analysis to determine whether the dietary pattern was in compliance.

2.6. Structural Equation Modelling

Structural equation modelling (SEqM), also known as covariance structural modelling, was developed by Joreskog in the 1970s [19]. SEqM consists of factor analysis and path analysis. Using this method, the acceptability of the theoretical model under specific factors can be tested. CFA was used to test the suitability of dietary patterns determined by EFA. SEqM uses a similar approach to through-path analysis to investigate the structural relationships between latent variables, and the regression path coefficients reflect the degree of correlation between latent variables. In this study, we constructed the SEqM to investigate the relationship between dietary pattern and high blood glucose. Goodness-of-fit index (GFI), adjusted comparative fit index (ACFI) ≥ 0.90, parsimonious goodness-of-fit index (PGFI), parsimonious baseline fit index (PNFI) ≥ 0.50 and root mean square error of approximation (RMSEA) ≤ 0.08 were used to confirm the degree of model fit.

2.7. Statistical Analysis

The mean $\pm$ standard deviation (mean $\pm$ SD) of continuous variables and frequencies of categorical variables were used to represent the distribution of general characteristics. EFA was used to identify the dietary patterns of the individuals, and the factor scores were divided into quartiles for further analysis. CFA allowed further determination of the suitability of EFA. Multivariate logistic regression analysis was used to calculate the OR and 95% CI for high blood glucose in each quartile of the factor scores. Structural equation modelling was used to investigate the correlation and degree of correlation between dietary patterns obtained from factor analysis and high blood glucose.

SAS 9.4 (Cary, NC, USA), IBM SPSS 26.0 (Armonk, NY, USA) and Origin (2021, Northampton, MA, USA) data analysis and plotting software were used for data management and statistical analysis. All statistical tests were two-sided, and differences were considered statistically significant when $p < 0.05$.

3. Results

3.1. Basic Information of Participants

Table 1 shows the basic information for the different gender groups. After exclusion, 3137 participants (54.4% male, $n = 1708$) with complete data were included in this study. The average age and energy intake of men were significantly higher than that of women ($p < 0.001$). The mean blood glucose level was higher in males than females (male = 5.5 $\pm$ 1.4 mmol/L vs. female = 5.4 $\pm$ 1.3 mmol/L, $p < 0.05$). The prevalence of high blood glucose in men was not significantly different compared to women (male = 9.3% vs. female = 8.1%, $p > 0.05$). Among participants, 45.7% ($n = 653$) males had smoking behaviour, which was significantly higher than females (1.4%) ($p < 0.001$).

3.2. Determination of Dietary Patterns

Figure 2 and Supplementary Table S2 illustrate the dietary patterns identified by the EFA. Factor loading and interpretability were used to explain the two dietary patterns generated, namely the "modern dietary pattern" (Pattern I) and the "fruit and milk dietary pattern". Then, we put food groupings with higher factor loadings from these two dietary patterns into the CFA model (Figure 3). Ultimately, we found that the modern dietary pattern was dominated by red meats and their products, fresh vegetables, seafood, condiments, whole cereals, main cereals, poultry, tubers and starches and their products and fungi and algae. The fruit–milk dietary pattern was dominated by nuts and seeds, fruits, eggs, milk and its products, pastry snacks and vegetable oils.

Table 1. Characteristics of the basic information distribution by gender.

Groups	Male		Female		*p*-Value
	Mean	**SD**	**Mean**	**SD**	
Age (years)	56.5	14.5	54.3	14.6	<0.001
Blood glucose (mmol/L)	5.5	1.4	5.4	1.3	0.031
BMI (kg/m^2)	24.9	3.3	24.8	3.6	0.091
Energy intake (kcal/d)	1847.5	540.4	1540.8	440.9	<0.001
	n	%	*n*	%	*p*-value
Age group (years)					<0.001
18~34	143	10.0	214	12.5	
35~49	269	18.8	375	22.0	
50~64	542	37.9	671	39.3	
65~	475	33.2	448	26.2	
BMI level					<0.001
Thinness	22	1.5	38	2.2	
Normal	528	36.9	703	41.2	
Overweight	635	44.4	667	39.1	
Obesity	244	17.1	300	17.6	
Central obesity					0.574
No	865	60.5	1017	59.5	
Yes	564	39.5	691	40.5	
Smoking behaviour					<0.001
No	776	54.3	1684	98.6	
Yes	653	45.7	24	1.4	
Diabetes					0.223
No	1296	90.7	1570	91.9	
Yes	133	9.3	138	8.1	

3.3. Analysis of the Relationship between Dietary Patterns and High Blood Glucose by Multivariate Logistic Regression

Table 2 shows the association between dietary pattern and high blood glucose in Jiangsu adults by using multivariate logistic regression modelling; the results suggest that the high intake of modern dietary patterns increased adults' risk of high blood glucose (composed of Q1, Q3~Q4 OR = 1.566 and 1.561, 95% CI: 1.063~2.308 and 1.025~2.379, respectively, $p < 0.05$) and showed a trend toward elevation with increasing intake ($P_{trend} < 0.05$). However, the fruit–milk dietary pattern had no significant association with high blood glucose ($p > 0.05$).

Table 2. Odds ratios (95% confidence intervals) for high blood glucose across quartiles of dietary patterns.

Groups	OR	95% CI	*p*-Value	*p* for Trend
Modern Dietary Pattern				
Q1	1.000			0.021
Q2	1.441	(0.992~2.094)	0.055	
Q3	1.566	(1.063~2.308)	0.023	
Q4	1.561	(1.025~2.379)	0.038	
Fruit–milk dietary pattern				
Q1	1.000			0.232
Q2	0.998	(0.687~1.451)	0.992	
Q3	1.060	(0.734~1.531)	0.755	
Q4	1.269	(0.887~1.814)	0.192	

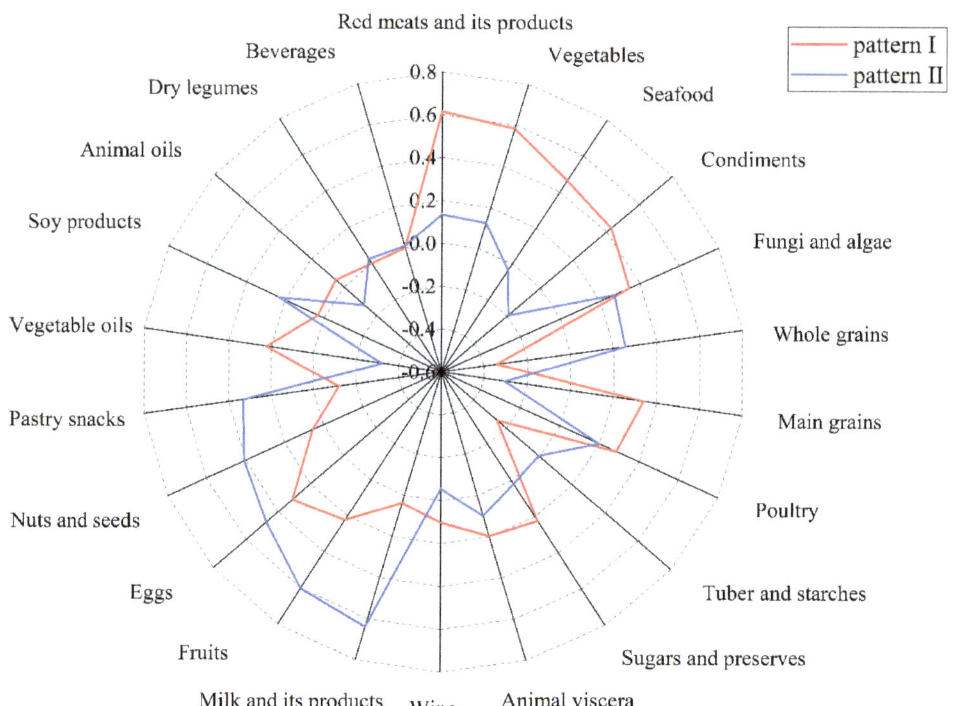

Figure 2. Radar plot of two dietary patterns by EFA.

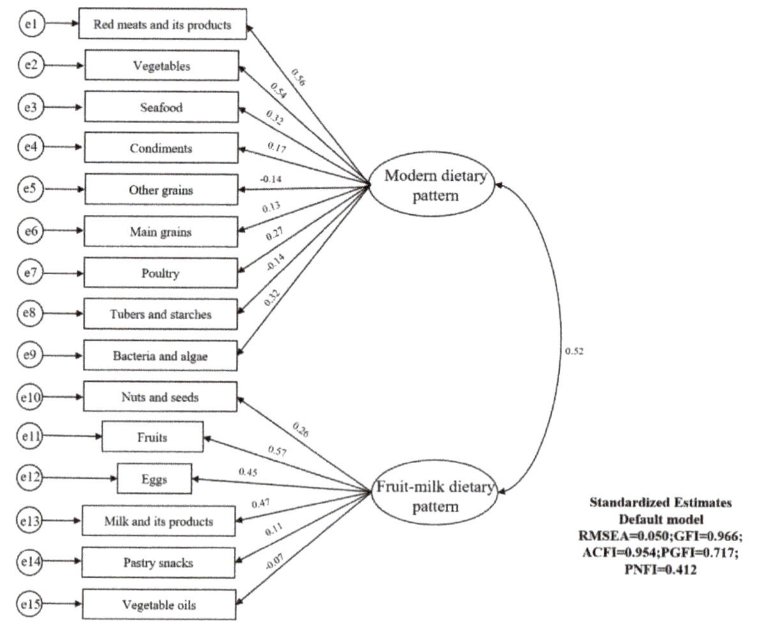

Figure 3. Measurement model for two dietary patterns by CFA. RMSEA = 0.050, GFI = 0.966, ACFI = 0.954, PGFI = 0.717 and PNFI = 0.412. e, error.

3.4. Structural Model

Figure 4 shows the SEqM plot with standardised estimates of the relationship between dietary patterns, demographic characteristics and high blood glucose. The final SEqM model was obtained by increasing residual correlations and modification indices (as shown in Figure 4 and Table 3). The goodness-of-fit indices of the final model indicated an acceptable fit (RMSEA = 0.068, GFI = 0.913, ACFI = 0.891, PGFI = 0.727). The modern dietary pattern was positively associated with the risk of high blood glucose among adults in Jiangsu province of China. ($\beta = 0.127$, $p < 0.001$).

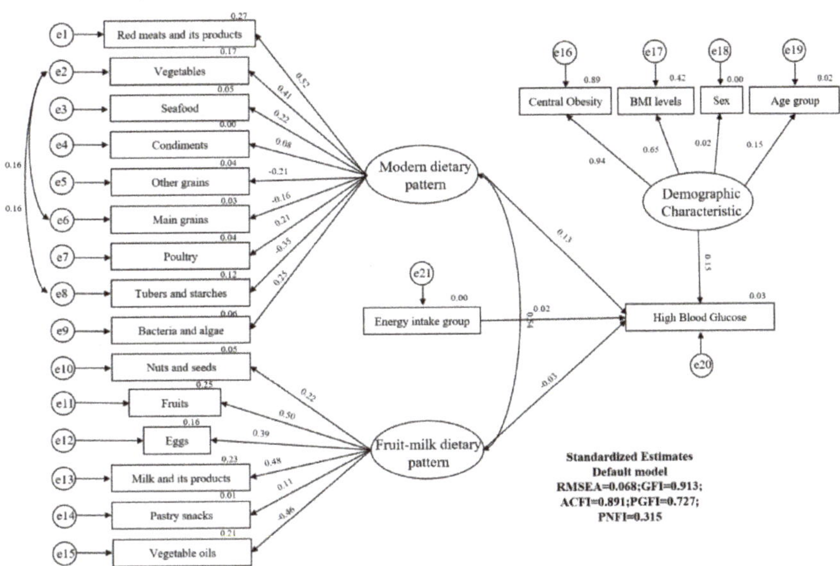

Figure 4. Final structural models. The path standardised coefficients of variables are presented on pathways. RMSEA = 0.068, GFI =0.913, ACFI = 0.891, PGFI= 0.727 and PNFI = 0.315. e, error.

Table 3. Parameter Estimates from the SEqM of dietary patterns and high blood glucose among adults.

Path Analysis	Non-Standardised Coefficient	Standardised Coefficients	S.E.	C.R.	*p*-Value
Modern dietary pattern →diabetes	0.001	0.127	0.000	3.417	<0.001
Fruit-milk dietary pattern→diabetes	−0.003	−0.032	0.003	−0.903	0.366

4. Discussion

Diabetes mellitus is one of the most prominent risk factors affecting the health of the population [20]. The risk of high blood glucose among adults is on the rise in Jiangsu province, China [21]. It is suggested that we should be on alert to the further risk of high blood glucose in the region. Diet, as a controllable lifestyle, has been shown to significantly influence the development of DM [22]. In our study, two dietary patterns were identified by exploratory factor analysis and confirmatory factor analysis: modern dietary pattern and fruit–milk dietary pattern. Multivariate logistic regression and SEqM analysis revealed that the modern dietary pattern was positively associated with high blood glucose among adults in Jiangsu province, China, while the fruit–milk dietary pattern was not significantly associated with high blood glucose.

The modern dietary pattern, which was rich in red meats and its products, vegetables, seafood, condiments, fungi and algae, main grains and poultry, but was low in whole grains and tubers and preserves and was significantly associated with high blood glucose in adults. It is similar to the modern dietary pattern obtained from the China Health and Nutrition Survey (CHNS) [23,24]. The survey showed that children and adolescents aged 6–14 years and the elderly aged 60 years and older who adhere to modern dietary patterns rich in saturated fat and cholesterol are at increased risk of obesity. Obesity, in turn, is an important risk factor for diabetes [25]. Besides, we believe that the positive association between modern dietary patterns and the risk of type 2 diabetes might be partly attributed to unhealthy dietary components, such as red meats and their products and main grains. First, red meat and processed meat products, which are rich in saturated fat, have been found to be significantly and positively associated with chronic diseases such as diabetes [26,27]. In this study, our analysis classified red meat and processed meat products as a food group. Their factor loadings were first in the modern dietary pattern, indicating a high intake of red and processed meat in people who prefer the modern dietary pattern. A meta-analysis found that consuming an additional 100 g of red meat per day increased the risk of developing T2DM, while consuming 50 g and more of processed meat products per day increased the risk of T2DM by 30% [10]. Excessive intake of red meat products may lead to the overabsorption of heme iron [28]. Internal iron overload may promote insulin resistance and increase the risk of T2DM [29]. Second, excessive intake of main grains and low intake of whole grains is another major feature of this dietary pattern. In our study, main grains refer to refined rice products and wheat products. There is a general consensus that people with or at risk of type 2 diabetes should avoid carbohydrate-rich foods [30]. A systematic review and meta-analysis indicated that consuming 200–400 g of refined grains per day may increase the risk of T2DM by 6−14% [10]. In another one of our studies, we also found that excess intake of refined carbohydrates could promote elevated blood glucose [31]. Meanwhile, it is important to pay attention to the quantity of carbohydrates as well as their source and quality. Numerous studies have found that less processed whole grain foods could improve blood glucose measurements in adults with type 2 diabetes more than the same number of refined grains [32,33]. The potential reason for it is that whole grains are likely to be digested by microbiota in the colon into short-chain fatty acids, which are absorbed without altering circulating blood glucose levels [34,35]. Results from the latest prospective cohort study also showed that participants with high whole grain intakes had a 29% lower incidence of type 2 diabetes, suggesting further support for the current recommendation to increase whole grain intake as part of a healthy diet to prevent type 2 diabetes [36]. Higher intake of vegetables was also a key component of this dietary pattern. Results from the Guangzhou Nutrition and Health Study (GNHS) showed that vegetables and gut microbiota diversity and composition were not associated with the risk of developing T2DM [37]. It may be related to the Chinese food culture, where most vegetables are cooked before consumption. One study showed that a higher intake of raw vegetables (rather than cooked vegetables) was positively associated with a lower risk of CVD [38]. Furthermore, the results of meta-analyses showed that a total vegetable intake of about 100 g per day did not affect the risk of developing T2DM. Increasing intake to 300 g per day could reduce the risk of developing T2DM by 9% with a non-linear dose-response association, whereas above this value, no significant benefit of increasing intake was observed [10,39]. Therefore, it is significant to investigate the potential different associations of raw vegetables and cooked vegetables and different intakes with T2DM in future work. Furthermore, high intakes of other meat products such as fish and poultry were also observed in this dietary pattern. However, a meta-analysis of prospective cohort studies found that these two food groups did not appear to be strongly associated with diabetes [10,40]. Finally, the intake of condiments such as salt was considered in the factor analysis and represented modern dietary patterns with a high factor loading. The proinflammatory response has an essential effect on the development of T2DM [41]. It has been shown that increased salt (sodium chloride) intake appeared to affect T2DM by enhancing TH17 cell activity through the p38/MAPK pathway and serum/glucocorticoid-regulated kinase 1 (SGK1) to increase proinflammatory cytokine levels [42]. In addition, a Japanese cohort study found that high HbA1C and dietary sodium intake had a synergistic effect, which increased the risk of CVD in patients with T2DM [43].

The fruit–milk dietary pattern, which is high in milk and its products, fruits, eggs, nuts and seeds and pastry snacks, but is low in vegetable oils, had no association with high blood glucose. This nonsignificant relationship might be the result of the interaction of certain healthy and unhealthy food groups. On the one hand, milk and its products, fruits and nuts and seeds as healthy foods may reduce the risk of developing DM. A prospective study from Singapore showed a significant 12% reduction in the risk of T2DM in daily milk drinkers compared to non-milk drinkers [44]. A meta-analysis including 14 cohort studies found a non-linear negative association of total dairy and low-fat dairy consumption with T2DM risk, with the inverse association appearing to be strongest at an intake of 200 g/day [39,45]. As the intake increased to 400–600 g/day, the risk was reduced by 6%. There was no significant benefit for increasing intake above this value [10]. In addition, total fruits, nuts and seeds were negatively associated with the risk of developing T2DM [10,37,46,47]. On the other hand, a higher intake of eggs and pastry snacks and a lower intake of vegetable oils were thought to be positively associated with T2DM [10,39]. Although this dietary pattern did not show a significant association with high blood glucose in this study, we need to recognise the drawbacks of this dietary pattern. For example, the huge market for snack foods or ultra-processed foods in China could cause a dramatic shock to the traditional Chinese diet.

To our knowledge, this is the first study to combine SEqM and multivariate logistic regression to examine the association between dietary patterns and high blood glucose in Jiangsu province, China. Moreover, our population was sampled according to strict criteria, with the results being representative. However, the present investigation has some shortcomings. First, the cross-sectional study design naturally hinders the inference of causality. Second, the data of the diet was chosen from a 3-day, 24-h dietary recall and weighing method, and seasonal factors may have led to biased food choices; third, other confounding factors, such as physical activity and sleep duration, were not considered in this study.

5. Conclusions

This study finally identified two dietary patterns through EFA and CFA: the modern dietary pattern and the fruit–milk dietary pattern. The modern dietary pattern characterised by a high intake of red meats was positively associated with high blood glucose among adults in Jiangsu province of China, while the fruit–milk dietary pattern was not significantly associated with high blood glucose.

Supplementary Materials: The following supporting information can be downloaded at: https://www.mdpi.com/article/10.3390/nu14194111/s1, Table S1. Food Groupings Used in Factor Analysis, Table S2. Factor Loadings for three Dietary Patterns Derived from Factor Analysis.

Author Contributions: Y.W. made interpretation of data and drafted the work; W.X. collected all the samples and drafted the work; T.T., J.Z., Q.Z., D.P., D.X. and Y.L. conducted the experiments and analysed the data; G.S. helped with the experiments; Y.D. designed and supervised the work. All authors have read and agreed to the published version of the manuscript.

Funding: This research was funded by the Study of the National Prevention and Control of Major Chronic Non-Communicable Diseases from Major Project of National Key R&D Program (No. 2016YFC1305201) of China, Nanjing Municipal Health Science and Technology Development Special Fund project (No. YKK19123) and Jiangsu Provincial geriatric health scientific research Project: Dietary nutrition evaluation and application of intervention technology for sarcopenia in the elderly (No. LKM 2022005).

Institutional Review Board Statement: This study was conducted according to the guidelines laid down in the Declaration of Helsinki and all procedures involving research study participants were approved by the Ethics Committee of the Chinese Center for Disease Control and Prevention (approval number: 201519-B, 2014). Written informed consent was obtained from all subjects.

Informed Consent Statement: Not applicable.

Data Availability Statement: The datasets used and/or analyzed during the current study are available from the corresponding author on reasonable request.

Acknowledgments: The authors thank all those who contributed to the physical examination, biochemical and nutritional assessment, and database management.

Conflicts of Interest: The authors declare no conflict of interest.

References

1. Cole, J.B.; Florez, J.C. Genetics of diabetes mellitus and diabetes complications. *Nat. Rev. Nephrol.* **2020**, *16*, 377–390. [CrossRef] [PubMed]
2. Haas, A.V.; McDonnell, M.E. Pathogenesis of Cardiovascular Disease in Diabetes. *Endocrinol. Metab. Clin. N. Am.* **2018**, *47*, 51–63. [CrossRef] [PubMed]
3. Eid, S.; Sas, K.M.; Abcouwer, S.F.; Feldman, E.L.; Gardner, T.W.; Pennathur, S.; Fort, P.E. New insights into the mechanisms of diabetic complications: Role of lipids and lipid metabolism. *Diabetologia* **2019**, *62*, 1539–1549. [CrossRef] [PubMed]
4. Goldstein, A.S.; Janson, B.J.; Skeie, J.M.; Ling, J.J.; Greiner, M.A. The effects of diabetes mellitus on the corneal endothelium: A review. *Surv. Ophthalmol.* **2020**, *65*, 438–450. [CrossRef] [PubMed]
5. Sun, H.; Saeedi, P.; Karuranga, S.; Pinkepank, M.; Ogurtsova, K.; Duncan, B.B.; Stein, C.; Basit, A.; Chan, J.C.N.; Mbanya, J.C.; et al. Christian Bommer Diabetes Atlas: Global, regional and country-level diabetes prevalence estimates for 2021 and projections for 2045. *Diabetes Res. Clin. Pract.* **2022**, *183*, 109119. [CrossRef] [PubMed]
6. China. *Report on the Status of Nutrition and Chronic Diseases Among Chinese Residents 2020*; People's Publishing House: Beijing, China, 2020.
7. Zaroudi, M.; Charati, J.Y.; Mehrabi, S.; Ghorbani, E.; Norouzkhani, J.; Shirashiani, H.; Nikzad, B.; Seiedpour, M.; Izadi, M.; Mirzaei, M.; et al. Dietary Patterns Are Associated with Risk of Diabetes Type 2: A Population-Based Case-Control Study. *Arch. Iran. Med.* **2016**, *19*, 166–172.
8. Ojo, O. Dietary Intake and Type 2 Diabetes. *Nutrients* **2019**, *11*, 2177. [CrossRef]
9. Ozcariz, S.G.; Bernardo Cde, O.; Cembranel, F.; Peres, M.A.; González-Chica, D.A. Dietary practices among individuals with diabetes and hypertension are similar to those of healthy people: A population-based study. *BMC Public Health* **2015**, *15*, 479. [CrossRef]
10. Schwingshackl, L.; Hoffmann, G.; Lampousi, A.-M.; Knüppel, S.; Iqbal, K.; Schwedhelm, C.; Bechthold, A.; Schlesinger, S.; Boeing, H. Food groups and risk of type 2 diabetes mellitus: A systematic review and meta-analysis of prospective studies. *Eur. J. Epidemiol.* **2017**, *32*, 363–375. [CrossRef]
11. Zhang, Y.; Tan, H.; Tang, J.; Li, J.; Chong, W.; Hai, Y.; Feng, Y.; Lunsford, L.D.; Xu, P.; Jia, D.; et al. Effects of Vitamin D Supplementation on Prevention of Type 2 Diabetes in Patients with Prediabetes: A Systematic Review and Meta-analysis. *Diabetes Care* **2020**, *43*, 1650–1658. [CrossRef]
12. Sprake, E.F.; Russell, J.M.; Cecil, J.E.; Cooper, R.J.; Grabowski, P.; Pourshahidi, L.K.; Barker, M.E. Dietary patterns of university students in the UK: A cross-sectional study. *Nutr. J.* **2018**, *17*, 90. [CrossRef]
13. O'Hara, C.; Gibney, E.R. Meal Pattern Analysis in Nutritional Science: Recent Methods and Findings. *Adv. Nutr.* **2021**, *12*, 1365–1378. [CrossRef]
14. Iqbal, K.; Schwingshackl, L.; Floegel, A.; Schwedhelm, C.; Stelmach-Mardas, M.; Wittenbecher, C.; Galbete, C.; Knüppel, S.; Schulze, M.B.; Boeing, H. Graphical models identified food intake networks and risk of type 2 diabetes, CVD, and cancer in the EPIC-Potsdam study. *Eur. J. Nutr.* **2019**, *58*, 1673–1686. [CrossRef]
15. Archundia Herrera, M.C.; Subhan, F.B.; Chan, C.B. Dietary Patterns and Cardiovascular Disease Risk in People with Type 2 Diabetes. *Curr. Obes. Rep.* **2017**, *6*, 405–413. [CrossRef]
16. Schulz, C.A.; Oluwagbemigun, K.; Nöthlings, U. Advances in dietary pattern analysis in nutritional epidemiology. *Eur. J. Nutr.* **2021**, *60*, 4115–4130. [CrossRef]
17. Castro, M.A.; Baltar, V.T.; Marchioni, D.M.; Fisberg, R.M. Examining associations between dietary patterns and metabolic CVD risk factors: A novel use of structural equation modelling. *Br. J. Nutr.* **2016**, *115*, 1586–1597. [CrossRef]
18. Tod, D.; Edwards, C.; Hall, G. Drive for leanness and health-related behavior within a social/cultural perspective. *Body Image* **2013**, *10*, 640–643. [CrossRef]
19. Jöreskog, K.G. Structural analysis of covariance and correlation matrices. *Psychometrika* **1978**, *43*, 443–477. [CrossRef]
20. Li, Y.; Teng, D.; Shi, X.; Qin, G.; Qin, Y.; Quan, H.; Shi, B.; Sun, H.; Ba, J.; Chen, B.; et al. Prevalence of diabetes recorded in mainland China using 2018 diagnostic criteria from the American Diabetes Association: National cross sectional study. *BMJ* **2020**, *369*, m997. [CrossRef]

21. Yue, J.; Mao, X.; Xu, K.; Lü, L.; Liu, S.; Chen, F.; Wang, J. Awareness, Treatment and Control of Diabetes Mellitus in a Chinese Population. *PLoS ONE* **2016**, *11*, e0153791. [CrossRef]
22. Papamichou, D.; Panagiotakos, D.B.; Itsiopoulos, C. Dietary patterns and management of type 2 diabetes: A systematic review of randomised clinical trials. *Nutr. Metab. Cardiovasc. Dis.* **2019**, *29*, 531–543. [CrossRef]
23. Zhen, S.; Ma, Y.; Zhao, Z.; Yang, X.; Wen, D. Dietary pattern is associated with obesity in Chinese children and adolescents: Data from China Health and Nutrition Survey (CHNS). *Nutr. J.* **2018**, *17*, 68. [CrossRef]
24. Xu, X.; Hall, J.; Byles, J.; Shi, Z. Dietary Pattern Is Associated with Obesity in Older People in China: Data from China Health and Nutrition Survey (CHNS). *Nutrients* **2015**, *7*, 8170–8188. [CrossRef]
25. Riaz, H.; Khan, M.S.; Siddiqi, T.J.; Usman, M.S.; Shah, N.; Goyal, A.; Khan, S.S.; Mookadam, F.; Krasuski, R.A.; Ahmed, H. Association Between Obesity and Cardiovascular Outcomes: A Systematic Review and Meta-analysis of Mendelian Randomization Studies. *JAMA Netw. Open.* **2018**, *1*, e183788. [CrossRef]
26. Wolk, A. Potential health hazards of eating red meat. *J. Intern. Med.* **2017**, *281*, 106–122. [CrossRef]
27. Talaei, M.; Wang, Y.L.; Yuan, J.M.; Pan, A.; Koh, W.P. Meat, Dietary Heme Iron, and Risk of Type 2 Diabetes Mellitus: The Singapore Chinese Health Study. *Am. J. Epidemiol.* **2017**, *186*, 824–833. [CrossRef]
28. Maghsoudi, Z.; Ghiasvand, R.; Salehi-Abargouei, A. Empirically derived dietary patterns and incident type 2 diabetes mellitus: A systematic review and meta-analysis on prospective observational studies. *Public Health Nutr.* **2016**, *19*, 230–241. [CrossRef]
29. Jiang, R.; Ma, J.; Ascherio, A.; Stampfer, M.J.; Willett, W.C.; Hu, F.B. Dietary iron intake and blood donations in relation to risk of type 2 diabetes in men: A prospective cohort study. *Am. J. Clin. Nutr.* **2004**, *79*, 70–75. [CrossRef] [PubMed]
30. McMacken, M.; Shah, S. A plant-based diet for the prevention and treatment of type 2 diabetes. *J. Geriatr. Cardiol.* **2017**, *14*, 342–354. [PubMed]
31. Wang, Y.-Y.; Zhang, J.-X.; Tian, T.; Gao, M.-Y.; Zhu, Q.-R.; Xie, W.; Fu, L.-M.; Wang, S.-K.; Dai, Y. Dietary patterns in association with the risk of elevated blood pressure, lipid profile and fasting plasma glucose among adults in Jiangsu Province of China. *Nutr. Metab. Cardiovasc. Dis.* **2022**, *32*, 69–79. [CrossRef] [PubMed]
32. Åberg, S.; Mann, J.; Neumann, S.; Ross, A.B.; Reynolds, A.N. Whole-Grain Processing and Glycemic Control in Type 2 Diabetes: A Randomized Crossover Trial. *Diabetes Care* **2020**, *43*, 1717–1723. [CrossRef]
33. Malin, S.K.; Kullman, E.L.; Scelsi, A.R.; Haus, J.M.; Filion, J.; Pagadala, M.R.; Jean-Philippe, G.; Kochhar, S.; Ross, A.B.; Kirwan, J.P. A whole-grain diet reduces peripheral insulin resistance and improves glucose kinetics in obese adults: A randomized-controlled trial. *Metabolism* **2018**, *82*, 111–117. [CrossRef]
34. Stephen, A.M.; Cummings, J.H. Mechanism of action of dietary fibre in the human colon. *Nature* **1980**, *284*, 283–284. [CrossRef]
35. Barber, T.M.; Kabisch, S.; Pfeiffer, A.F.H.; Weickert, M.O. The Health Benefits of Dietary Fibre. *Nutrients* **2020**, *12*, 3209. [CrossRef]
36. Hu, Y.; Ding, M.; Sampson, L.; Willett, W.C.; Manson, J.E.; Wang, M.; Rosner, B.; Hu, F.B.; Sun, Q. Intake of whole grain foods and risk of type 2 diabetes: Results from three prospective cohort studies. *BMJ* **2020**, *370*, m2206. [CrossRef]
37. Jiang, Z.; Sun, T.-Y.; He, Y.; Gou, W.; Zuo, L.-S.; Fu, Y.; Miao, Z.; Shuai, M.; Xu, F.; Xiao, C.; et al. Dietary fruit and vegetable intake, gut microbiota, and type 2 diabetes: Results from two large human cohort studies. *BMC Med.* **2020**, *18*, 371. [CrossRef]
38. Feng, Q.; Kim, J.H.; Omiyale, W.; Bešević, J.; Conroy, M.; May, M.; Yang, Z.; Wong, S.Y.; Tsoi, K.K.; Allen, N.; et al. Raw and Cooked Vegetable Consumption and Risk of Cardiovascular Disease: A Study of 400,000 Adults in UK Biobank. *Front. Nutr.* **2022**, *9*, 831470. [CrossRef]
39. Neuenschwander, M.; Ballon, A.; Weber, K.S.; Norat, T.; Aune, D.; Schwingshackl, L.; Schlesinger, S. Role of diet in type 2 diabetes incidence: Umbrella review of meta-analyses of prospective observational studies. *BMJ* **2019**, *366*, l2368. [CrossRef]
40. Yang, X.; Li, Y.; Wang, C.; Mao, Z.; Zhou, W.; Zhang, L.; Fan, M.; Cui, S.; Li, L. Meat and fish intake and type 2 diabetes: Dose-response meta-analysis of prospective cohort studies. *Diabetes Metab.* **2020**, *46*, 345–352. [CrossRef]
41. Cruz, N.G.; Sousa, L.P.; Sousa, M.O.; Pietrani, N.T.; Fernandes, A.P.; Gomes, K.B. The linkage between inflammation and Type 2 diabetes mellitus. *Diabetes Res. Clin. Pract.* **2013**, *99*, 85–92. [CrossRef]
42. Kleinewietfeld, M.; Manzel, A.; Titze, J.; Kvakan, H.; Yosef, N.; Linker, R.A.; Muller, D.N.; Hafler, D.A. Sodium chloride drives autoimmune disease by the induction of pathogenic TH17 cells. *Nature* **2013**, *496*, 518–522. [CrossRef]
43. Horikawa, C.; Yoshimura, Y.; Kamada, C.; Tanaka, S.; Tanaka, S.; Hanyu, O.; Araki, A.; Ito, H.; Tanaka, A.; Ohashi, Y.; et al. Dietary sodium intake and incidence of diabetes complications in Japanese patients with type 2 diabetes: Analysis of the Japan Diabetes Complications Study (JDCS). *J. Clin. Endocrinol. Metab.* **2014**, *99*, 3635–3643. [CrossRef]
44. Talaei, M.; Pan, A.; Yuan, J.M.; Koh, W.P. Dairy intake and risk of type 2 diabetes. *Clin. Nutr.* **2018**, *37*, 712–718. [CrossRef]
45. Gao, D.; Ning, N.; Wang, C.; Wang, Y.; Li, Q.; Meng, Z.; Liu, Y.; Li, Q. Dairy products consumption and risk of type 2 diabetes: Systematic review and dose-response meta-analysis. *PLoS ONE* **2013**, *8*, e73965. [CrossRef]
46. Yu, Z.; Malik, V.S.; Keum, N.; Hu, F.B.; Giovannucci, E.L.; Stampfer, M.J.; Willett, W.C.; Fuchs, C.S.; Bao, Y. Associations between nut consumption and inflammatory biomarkers. *Am. J. Clin. Nutr.* **2016**, *104*, 722–728. [CrossRef]
47. Muraki, I.; Imamura, F.; Manson, E.J.; Hu, F.B.; Willett, W.C.; van Dam, R.; Sun, Q. Fruit consumption and risk of type 2 diabetes: Results from three prospective longitudinal cohort studies. *BMJ* **2013**, *347*, f5001. [CrossRef]

 nutrients

Article

Inverse Association between Oxidative Balance Score and Incident Type 2 Diabetes Mellitus

Yu-Jin Kwon [1,†], Hye-Min Park [2,†] and Jun-Hyuk Lee [3,4,*]

1 Department of Family Medicine, Yongin Severance Hospital, Yonsei University College of Medicine, Yongin 16995, Republic of Korea; digda3@yuhs.ac
2 Primary Care Research Center, National Health Insurance Service, Ilsan Hospital, Goyang 10326, Republic of Korea; jadorehm@nhimc.or.kr
3 Department of Family Medicine, Nowon Eulji Medical Center, Eulji University School of Medicine, Seoul 01830, Republic of Korea
4 Department of Medicine, School of Medicine, Hanyang University, Seoul 04763, Republic of Korea
* Correspondence: swpapa@eulji.ac.kr; Tel.: +82-2-970-8515 or +82-10-8687-4780; Fax: +82-2-970-8862
† These authors contributed equally to this work.

Abstract: Mitigating the risk of type 2 diabetes mellitus (T2DM) can be achieved through the maintenance of a healthy weight, the adoption of a healthy diet, and engaging in regular physical activity. The oxidative balance score (OBS), an integrated measure of pro- and antioxidant exposure conditions, represents an individual's overall oxidative balance status. This study aimed to evaluate the association between OBS and T2DM incidence using data from a large, community-based, prospective cohort study. Data from 7369 participants aged 40–69 years who engaged in the Korean Genome and Epidemiology Study (KoGES) were analyzed. The hazard ratio (HR) and 95% confidence interval (CI) for T2DM incidence of sex-specific OBS tertile groups were calculated using univariable and multivariable Cox proportional hazard regression analyses. During the mean 13.6-year follow-up period, 908 men and 880 women developed T2DM. The fully-adjusted HR (95% CI) for incident T2DM of the middle and highest tertile groups, compared with the referent lowest tertile group, were 0.86 (0.77–1.02) and 0.83 (0.70–0.99) in men and were 0.94 (0.80–1.11) and 0.78 (0.65–0.94) in women, respectively. Individuals with a high OBS are at lower risk for the development of T2DM. This implies that lifestyle modification with more antioxidant properties could be a preventive strategy for T2DM.

Keywords: oxidative balance score; antioxidant; type 2 diabetes mellitus; Korean genome; epidemiology study

Citation: Kwon, Y.-J.; Park, H.-M.; Lee, J.-H. Inverse Association between Oxidative Balance Score and Incident Type 2 Diabetes Mellitus. *Nutrients* **2023**, *15*, 2497. https://doi.org/10.3390/nu15112497

Academic Editor: Giuseppe Della Pepa

Received: 27 April 2023
Revised: 21 May 2023
Accepted: 22 May 2023
Published: 27 May 2023

Copyright: © 2023 by the authors. Licensee MDPI, Basel, Switzerland. This article is an open access article distributed under the terms and conditions of the Creative Commons Attribution (CC BY) license (https://creativecommons.org/licenses/by/4.0/).

1. Introduction

Globally, the number of persons with diabetes mellitus is rising, with the International Diabetes Federation estimating that there were 463 million cases of the disease in 2019 and that there would be 700.2 million cases by 2045 [1]. The prevalence of type 2 diabetes mellitus (T2DM) among South Korean adults rapidly increased over the past decades from 1.5% in 1971 to 13.7% in 2016 [2]. The economic burden of diabetes mellitus in Korea was USD 18,293 million in 2019 [3]. Moreover, the per capita cost increased nearly four times, from USD 3991 to USD 11,965, when the number of complications due to diabetes mellitus increased from one to three or more [3]. Patients with T2DM are at a higher risk of cardiovascular mortality, all-cause mortality, and comorbidities, including cardiovascular disease, cerebrovascular disease, and peripheral vascular disease [4]. Therefore, preventive strategies for T2DM have been emphasized to reduce this disease burden [4].

The two main characteristics of T2DM are target tissue insulin resistance and a relative deficiency of insulin production from pancreatic β-cells [5]. Over recent years, numerous studies have demonstrated a synergistic interaction between inflammation-related insulin resistance [6]. The emerging role of chronic low-grade inflammation in insulin resistance and β-cell dysfunction in T2DM has engendered increasing attention in targeting inflammation to advance the prevention and management of the disease [7].

A recent meta-analysis elucidated that a coalescence of low-risk lifestyle behaviors (such as appropriate body weight, healthy eating habits, light alcohol consumption, regular exercise, and smoking cessation) resulted in an 80% reduction in the risk of developing T2DM [8]. This finding aligns with a previous study emphasizing the balance between antioxidants and oxidative stress in chronic diseases [9]. Smoking is a powerful pro-oxidant, and the burden of oxidative stress could be exacerbated through the secondary release of oxygen radicals from inflammation status [9]. Therefore, several studies proposed a link between chronic disease and the oxidative balance score (OBS) [10–17]. The OBS evaluates the oxidative balance of the lifestyle pattern of a subject in terms of the incorporated consumption of anti- and pro-oxidants [10,12,18]. Lifestyle (cigarette smoking and alcohol drinking), healthy body weight (obesity and abdominal obesity), and healthy diet (lower intakes of saturated fatty acid [SFA], omega-6 poly-unsaturated fatty acid [PUFA], iron and high intakes of vitamin C, vitamin E, omega-3 PUFA, selenium, and beta-carotene) could be involved as OBS components [10].

To the best of our knowledge, no previous study has comprehensively examined OBS and the incidence of T2DM in the middle-aged and elderly. Therefore, we prospectively investigated the development of T2DM according to the OBSs of tertile groups from a large-population, community-based Korean cohort observed over 16 years.

2. Materials and Methods

2.1. Study Population

We used the Korean Genome and Epidemiology Study (KoGES)-Ansan and Ansung, embedded in the KoGES, a large, community-based study in Korea. The study design and procedures were detailed in a previous study [19]. KoGES-Ansan and Ansung included 10,030 adults aged from 40 to 69 years. This survey was conducted between 2001 and 2002 and followed up every two years. For the present study, we included participants with the eighth follow-up, conducted between 2017 and 2018. Initially, a total of 10,030 participants received a health examination and questionnaire. Subsequently, we excluded participants with T2DM at baseline (n = 1351), those with missing data to evaluate T2DM (n = 2), those with missing data to calculate OBSs (n = 601), and those who did not follow up after the baseline survey (n = 707). Finally, a total of 7369 participants (3485 men and 3884 women) were included in this analysis. The flow chart is shown in Figure 1. All participants in this study provided informed consent. This study protocol was approved by the Nowon Eulji Medical Center's Institutional Review Board (approval number: 2021-09-025) and followed the ethical criteria of the 1964 Declaration of Helsinki and its subsequent amendments.

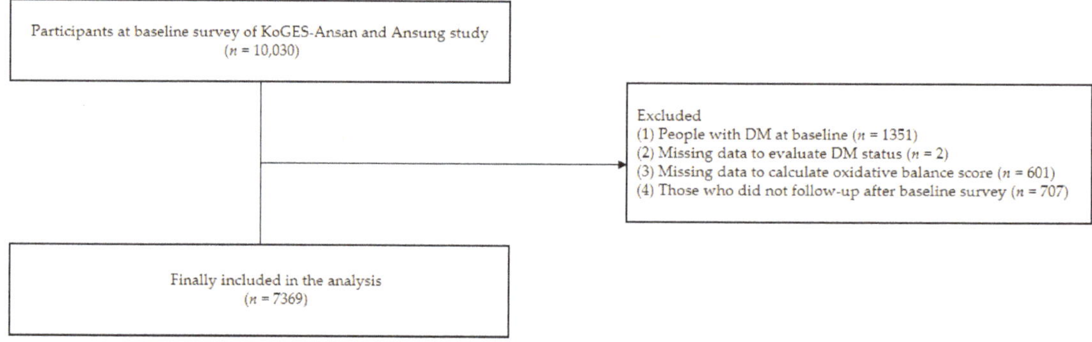

Figure 1. Flowchart of the study population selection.

2.2. Assessment of Oxidative Balance Score

The OBS was calculated as the sum of seven pro-oxidant factors and six antioxidant factors selected based on previous studies [10–17]. The scheme of OBS is described in Table 1. Pro-oxidant factors include SFA, omega-6 PUFA, total iron intake, smoking status, drinking status, obesity status, and abdominal obesity status. Each question was scored 0, 1, or 2. The scores for SFA, omega-6 PUFA, and total iron intake were assigned 0 through 2 points according to the sex-specific tertile values of each variable corresponding to low (score 2), intermediate (score 1), and high (score 0). For smoking status, the scores for never smoker, former smoker, and current smoker were 2, 1, and 0, respectively. For drinking status, the scores for a non-drinker, mild drinker (1–19 g/day in men, 1–9 g/day in women), and moderate drinker (20–29 g/day in men, 10–19 g/day in women) were 2, 1, and 0, respectively. Zero points were given for people with obesity, one point was given for people who were overweight, and 2 points were given for people within a normal weight range. Zero points were given for people with abdominal obesity. Antioxidant factors include intakes of omega-3 PUFA, vitamin C, vitamin E, selenium, and beta-carotene and physical activity. The scores for omega-3 PUFA, vitamin C, vitamin E, selenium, and beta-carotene intake were assigned 0 through 2 points according to the sex-specific tertile values of each variable corresponding to high (score 2), intermediate (score 1), and low (score 0). Two points were given for high-intensity physical activity, one for moderate physical activity, and 0 for low physical activity. The sums of the OBSs ranged from 0 to 26 points. We classified the participants into sex-specific tertile groups according to OBSs.

Table 1. Oxidative balance score assignment scheme.

OBS Components	Assignment Scheme *
1. Saturated fatty acid [P]	0 = high (3rd tertile), 1 = intermediate (2nd tertile), 2 = low (1st tertile)
2. Omega-6 PUFA intake [P]	0 = high (3rd tertile), 1 = intermediate (2nd tertile), 2 = low (1st tertile)
3. Total iron intake [P]	0 = high (3rd tertile), 1 = intermediate (2nd tertile), 2 = low (1st tertile)
4. Smoking status [P]	2 = never smoker, 1 = former smoker, 0 = current smoker
5. Drinking status [P]	2 = non-drinker, 1 = mild-to-moderate drinker (<30 g/day in men, <20 g/day in women), 0 = heavy drinker (≥30 g/day in men, ≥20 g/day in women)
6. Overweight/obese [P]	2 = normal, 1 = overweight, 0 = obese
7. Abdominal obesity [P]	1 = normal, 0 = abdominal obesity
8. Omega-3 PUFA intake [A]	0 = low (1st tertile), 1 = intermediate (2nd tertile), 2 = high (3rd tertile)
9. Vitamin C intake [A]	0 = low (1st tertile), 1 = intermediate (2nd tertile), 2 = high (3rd tertile)
10. Vitamin E intake [A]	0 = low (1st tertile), 1 = intermediate (2nd tertile), 2 = high (3rd tertile)
11. Selenium intake [A]	0 = low (1st tertile), 1 = intermediate (2nd tertile), 2 = high (3rd tertile)
12. Total beta-carotene intake [A]	0 = low (1st tertile), 1 = intermediate (2nd tertile), 2 = high (3rd tertile)
13. Physical activity [A]	0 = low (<7.5 METs-h/wk), 1 = moderate (7.5–30 METs-h/wk), 2 = high (>30 METs-h/wk)

* Low, intermediate, and high categories correspond to sex-specific tertile values among participants in the KoGES at the baseline survey. Abbreviations: P, pro-oxidant; A, antioxidant; PUFA, poly-unsaturated fatty acid; MET, metabolic equivalent of task; KoGES, Korean Genome and Epidemiology Study.

2.3. Assessment of T2DM

T2DM was characterized as the presence of one or more of the following criteria: (1) a fasting plasma glucose level of 126 mg/dL or higher, (2) a 2 h after 75 g oral glucose tolerance test plasma glucose level of 200 mg/dL or higher, (3) glycosylated hemoglobin of 6.5% or higher, (4) treatment with oral anti-diabetic medicine, or (5) treatment with insulin therapy [20].

2.4. Covariates

A well-trained medical staff conducted health examinations and interviews according to a standard protocol. The detailed protocol of KoGES was available on the website (http://www.cdc.go.kr/contents.es?mid=a40504010000, accessed on 23 January 2023). Body mass index (BMI) was calculated as a person's weight in kilograms divided by the square of height in meters. Overweight was defined as when a person's BMI was

23 kg/m^2 or higher, and obesity was defined as when a person's BMI was 25 kg/m^2 or higher, respectively, based on the 2018 Korean Society for the Study of Obesity (KSSO) guideline [21]. Abdominal obesity was defined as a person's waist circumference (WC) being 90 cm or higher in men and 85 cm or higher in women, based on the 2018 KSSO guideline [21]. Mean blood pressure (MBP, mmHg) was calculated as diastolic blood pressure (DBP) + $1/3 \times$ [systolic blood pressures (SBP)-DBP]. Information about smoking, alcohol consumption, physical activity, education level, and household income was obtained from the self-reported questionnaires. A participant who had never smoked or smoked less than 100 cigarettes in their lifetime was defined as a never smoker. A participant who quit smoking and smoked more than 100 cigarettes during their lifetime was defined as a former smoker. A participant who smoked currently and had smoked more than 100 cigarettes during their lifetime was defined as a current smoker. We calculated each participant's daily alcohol intake (g/day). A heavy drinker was defined as a person who drinks alcohol more than 30 g/day in men and more than 20 g/day in women. A mild-to-moderate drinker was defined as a person who drinks alcohol below 30 g/day in men and below 20 g/day in women. A non-drinker was defined as a person who did not drink alcohol. Physical activity was measured as metabolic equivalent of task (MET)-hours per day (MET-h/day) using the International Physical Activity Questionnaire [22]. A nutrition survey was conducted through a face-to-face interview in an individual's home. Total energy intake and nutritional status were calculated using a validated 103-item food frequency questionnaire [23]. We used the daily total energy intake (kcal/day), omega-6 PUFA, total iron (mg/day), SFA (g/day), omega-3 PUFA (g/day), selenium (μg/day), vitamin C (mg/day), vitamin E (mg/day), and beta-carotene (μg/day) intake. The educational levels were classified as elementary/middle school, high school, and college/university. Monthly household income was categorized into less than 100 million South Korean Won, 100–200 million South Korean Won, and more than 200 million South Korean Won. The plasma glucose, serum insulin, total cholesterol, triglyceride, high-density lipoprotein (HDL) cholesterol, and C-reactive protein (CRP) were measured after at least 8 h of fasting using a Hitachi 700-110 Chemistry Analyzer (Hitachi, Ltd., Tokyo, Japan).

2.5. Statistical Analysis

After the normality test, variables with normal distribution were presented as mean $\pm$ standard deviations, and those with non-normal distribution were represented as median (25th, 75th). Continuous variables were compared using the one-way analysis of variance or using the Kruskal–Wallis test according to the sex-specific OBS tertiles. All statistical analyses were performed in a sex-specific manner. Categorical variables were represented as a number (%) and compared using the chi-square test. To determine cumulative incidence T2DM according to the sex-specific OBS tertiles, Kaplan–Meier curves with the log-rank test were utilized. We calculated the hazard ratio (HR) and 95% confidence interval (CI) for incident T2DM in the sex-specific middle tertile (T2) and highest tertile (T3) groups compared with the referent lowest tertile (T1) group using univariable and multivariable Cox proportional hazard regression analyses. We included age, total energy intake, MBP, education level, household income, fasting plasma glucose, serum insulin, serum total cholesterol, serum triglyceride, and serum CRP levels in the adjusted model. All statistical analyses were performed with SAS software (version 9.4; SAS Institute Inc., Cary, NC, USA) and R software (version 4.1.1; R Foundation for Statistical Computing, Vienna, Austria). A p-value less than 0.05 was regarded as statistically significant.

3. Results

3.1. Baseline Characteristics of the Study Population

Table 2 shows the baseline characteristics of the study population according to the OBS tertiles in men and women. In men, the T1 group had higher levels of MBP ($p = 0.001$), serum glucose ($p = 0.008$), insulin ($p < 0.001$), total cholesterol ($p < 0.001$), triglyceride ($p < 0.001$), CRP ($p < 0.001$), and total energy intake ($p < 0.001$) and had lower levels of HDL cholesterol ($p < 0.001$). In women, the T1 group had higher levels of MBP ($p < 0.001$), serum glucose ($p < 0.001$), insulin ($p < 0.001$), total cholesterol ($p < 0.001$), triglyceride ($p < 0.001$), CRP ($p < 0.001$), and total energy intake ($p < 0.001$) and had lower levels of HDL cholesterol ($p < 0.001$). The proportion of higher education level and household income was significantly higher in the T3 group in both men and women.

Table 2. Baseline characteristics of the study population.

Variables	Oxidative Balance Score							
	Men				Women			
	T1 (*n* = 1007)	T2 (*n* = 1060)	T3 (*n* = 1418)	*p* *	T1 (*n* = 1391)	T2 (*n* = 1212)	T3 (*n* = 1281)	*p* *
Age, years	50.8 ± 8.5	51.8 ± 8.7	51.1 ± 8.7	0.501	53.9 ± 9.0	52.1 ± 8.9	50.0 ± 8.3	<0.001
MBP, mmHg	98.2 ± 12.0	98.1 ± 12.3	96.5 ± 12.7	0.001	97.4 ± 13.8	95.2 ± 13.6	91.7 ± 13.0	<0.001
Glucose, mg/dL	85.1 ± 9.4	84.9 ± 9.1	84.1 ± 8.7	0.008	81.9 ± 8.0	81.0 ± 7.6	80.5 ± 7.4	<0.001
Insulin, µU/mL	6.8 [4.8; 9.5]	6.5 [4.9; 8.9]	6.1 [4.7; 8.3]	<0.001	7.7 [5.8; 10.2]	7.3 [5.5; 9.9]	7.1 [5.3; 9.3]	<0.001
Total cholesterol, mg/dL	193.8 ± 34.5	191.7 ± 35.3	188.8 ± 34.1	<0.001	192.6 ± 34.0	189.7 ± 35.1	185.0 ± 32.0	<0.001
Triglyceride, mg/dL	168.0 [124.0; 231.5]	141.5 [109.0; 200.5]	129.0 [95.0; 182.0]	<0.001	134.0 [100.0; 182.0]	120.0 [91.5; 163.0]	111.0 [87.0; 150.0]	<0.001
HDL cholesterol, mg/dL	42.7 ± 9.6	43.7 ± 9.5	44.6 ± 10.5	<0.001	44.4 ± 9.7	46.4 ± 10.2	47.1 ± 9.9	<0.001
CRP, mg/dL	0.16 [0.08; 0.27]	0.15 [0.07; 0.26]	0.13 [0.06; 0.22]	<0.001	0.14 [0.08; 0.24]	0.14 [0.06; 0.23]	0.11 [0.04; 0.20]	<0.001
Education level, *n* (%)				0.595				<0.001
Elementary/middle school	427 (42.4%)	451 (42.7%)	569 (40.3%)		1052 (76.3%)	797 (66.3%)	724 (56.7%)	
High school	366 (36.4%)	378 (35.8%)	514 (36.4%)		274 (19.9%)	317 (26.4%)	436 (34.2%)	
College/university	213 (21.2%)	226 (21.4%)	330 (23.4%)		53 (3.8%)	88 (7.3%)	116 (9.1%)	
Household income, *n* (%)				0.005				<0.001
<100 million South Korean Won	259 (25.9%)	272 (25.8%)	380 (26.9%)		658 (48.1%)	474 (39.6%)	414 (32.9%)	
100–200 million South Korean Won	338 (33.8%)	286 (27.1%)	447 (31.7%)		375 (27.4%)	361 (30.2%)	345 (27.4%)	
>200 million South Korean Won	402 (40.2%)	496 (47.1%)	581 (41.4%)		336 (24.5%)	362 (30.2%)	498 (39.6%)	
Energy intake, kcal/day	1798.5 ± 487.5	1979.5 ± 650.1	2250.6 ± 748.2	<0.001	1607.3 ± 513.4	1876.1 ± 656.9	2211.7 ± 858.5	<0.001

* *p*-value for the comparison of the baseline characteristics among sex-specific tertile groups of oxidative balance scores at the baseline survey. Significance was set at $p < 0.05$. Abbreviations: MBP, mean blood pressure; CRP, C-reactive protein.

Table 3 shows the baseline characteristics of individual components in relation to sex-specific OBS tertile groups. In both men and women, the T3 group had a higher SFA ($p < 0.001$), omega-6 PUFA ($p < 0.001$), total iron ($p < 0.001$), omega-3 PUFA ($p < 0.001$), vitamin C ($p < 0.001$), vitamin E ($p < 0.001$), selenium ($p < 0.001$), and beta-carotene intake ($p < 0.001$). In both men and women, those in the T1 group were more likely to be people with obesity ($p < 0.001$), current drinkers ($p < 0.001$), and current smokers ($p < 0.001$); had abdominal obesity ($p < 0.001$); and had lower physical activity ($p < 0.001$) than other groups.

Table 3. Individual components of the score by oxidative balance score tertiles.

Variables	Oxidative Balance Score							
	Men				Women			
	T1 (n = 1007)	T2 (n = 1060)	T3 (n = 1418)	p *	T1 (n = 1391)	T2 (n = 1212)	T3 (n = 1281)	p *
Saturated fatty acid, g/day	8.7 ± 4.0	10.5 ± 6.2	12.9 ± 7.7	<0.001	7.9 ± 4.5	10.5 ± 6.3	13.9 ± 9.6	<0.001
Omega-6 PUFA intake, g/day	7.6 ± 4.0	8.6 ± 5.5	9.2 ± 5.4	<0.001	7.5 ± 4.1	8.6 ± 5.6	9.8 ± 6.4	<0.001
Total iron intake, mg/day	16.2 ± 6.1	19.3 ± 8.8	23.5 ± 10.9	<0.001	14.6 ± 6.2	18.8 ± 8.5	24.5 ± 12.9	<0.001
Smoking status, n (%)				<0.001				<0.001
Current smoker	647 (64.3%)	519 (49.0%)	471 (33.2%)		86 (6.2%)	30 (2.5%)	7 (0.5%)	
Former smoker	273 (27.1%)	352 (33.2%)	442 (31.2%)		27 (1.9%)	12 (1.0%)	4 (0.3%)	
Never smoker	87 (8.6%)	189 (17.8%)	505 (35.6%)		1278 (91.9%)	1170 (96.5%)	1270 (99.1%)	
Drinking status, n (%)				<0.001				<0.001
Heavy drinker	295 (29.3%)	200 (18.9%)	167 (11.8%)		37 (2.7%)	12 (1.0%)	8 (0.6%)	
Mild-to-moderate drinker	565 (56.1%)	588 (55.5%)	680 (48.0%)		433 (31.1%)	315 (26.0%)	236 (18.4%)	
Non-drinker	147 (14.6%)	272 (25.7%)	571 (40.3%)		921 (66.2%)	885 (73.0%)	1037 (81.0%)	
Obesity status, n (%)				<0.001				<0.001
Obese	596 (59.2%)	426 (40.2%)	322 (22.7%)		892 (64.1%)	523 (43.2%)	267 (20.8%)	
Overweight	228 (22.6%)	301 (28.4%)	411 (29.0%)		345 (24.8%)	298 (24.6%)	384 (30.0%)	
Normal weight	183 (18.2%)	333 (31.4%)	685 (48.3%)		154 (11.1%)	391 (32.3%)	630 (49.2%)	
Abdominal obesity, n (%)	350 (34.8%)	226 (21.3%)	126 (8.9%)	<0.001	755 (54.3%)	374 (30.9%)	199 (15.5%)	<0.001
Omega-3 PUFA intake, g/day	1.1 ± 0.6	1.3 ± 0.9	1.5 ± 0.9	<0.001	1.0 ± 0.6	1.3 ± 0.9	1.6 ± 1.1	<0.001
Vitamin C intake, mg/day	73.8 ± 45.1	109.4 ± 85.9	160.0 ± 116.9	<0.001	81.5 ± 64.8	135.1 ± 116.3	205.6 ± 149.7	<0.001
Vitamin E intake, mg/day	10.5 ± 4.0	13.6 ± 6.6	17.8 ± 8.5	<0.001	9.2 ± 4.4	13.6 ± 7.1	19.2 ± 10.5	<0.001
Selenium intake, µg/day	36.9 ± 18.5	49.5 ± 31.0	66.2 ± 38.4	<0.001	29.6 ± 18.6	46.0 ± 28.6	67.1 ± 47.8	<0.001
Beta-carotene intake, µg/day	2212.2 ± 1530.4	3409.7 ± 2930.3	4938.1 ± 3940.7	<0.001	1980.8 ± 1415.3	3283.4 ± 2557.0	5325.7 ± 4545.6	<0.001
Physical activity, n (%)				<0.001				<0.001
Low (<7.5 METs-h/day)	103 (10.2%)	51 (4.8%)	47 (3.3%)		197 (14.2%)	97 (8.0%)	63 (4.9%)	
Moderate (7.5–30 METs-h/day)	682 (67.7%)	629 (59.3%)	774 (54.6%)		847 (60.9%)	787 (64.9%)	770 (60.1%)	
High (>30 METs-h/day)	222 (22.0%)	380 (35.8%)	597 (42.1%)		347 (24.9%)	328 (27.1%)	448 (35.0%)	

* p-value for the comparison of the baseline characteristics among sex-specific tertile groups of oxidative balance score at the baseline survey. Significance was set at $p < 0.05$. Abbreviations: PUFA, poly-unsaturated fatty acid; MET, metabolic equivalent of task.

3.2. Longitudinal Association of OBS and Incident T2DM

Throughout the average 13.6-year follow-up period, 908 (26.05%) men and 880 (22.66%) women developed new-onset T2DM.

Figure 2 presents the cumulative new-onset T2DM according to the sex-specific OBS tertiles as Kaplan–Meier curves. The T3 group showed the significantly lowest cumulative incident T2DM, followed by the T2 and T1 groups, in both men and women (both p-values for log-rank test < 0.001) (Figure 2a,b).

Table 4 shows the relationship between OBSs and incident T2DM in men and women. In men, the incidence rate per 1000 person-years was 27.49 in T1, 23.21 in T2, and 19.61 in T3. Compared with referent T1, the HR and 95% CI for new-onset T2DM were 0.85 (0.72–0.99) in T2 and 0.72 (0.62–0.85) in T3 (p for trend < 0.001). In the adjusted model, the HR and 95% CI for new-onset T2DM were 0.86 (0.73–1.02) in T2 and 0.83 (0.70–0.99) in T3 (p for trend = 0.035), compared with referent T1. In women, the incidence rate per 1000 person-years was 22.65 in T1, 19.09 in T2, and 14.48 in T3. The HR and 95% CI for new-onset T2DM were 0.84 (0.72–0.98) in T2 and 0.64 (0.54–0.75) in T3 (p for trend <0.001), compared with referent T1. The adjusted HR and 95% CI for new-onset T2DM were 0.94 (0.80–1.11) in T2 and 0.78 (0.65–0.94) in T3 (p for trend = 0.010), compared with referent T1. The HR and 95% CI for new-onset T2DM per one increment of OBS were 0.94 (0.91–0.96) in men and 0.91 (0.89–0.94) in women. Similar trends were shown in the adjusted model.

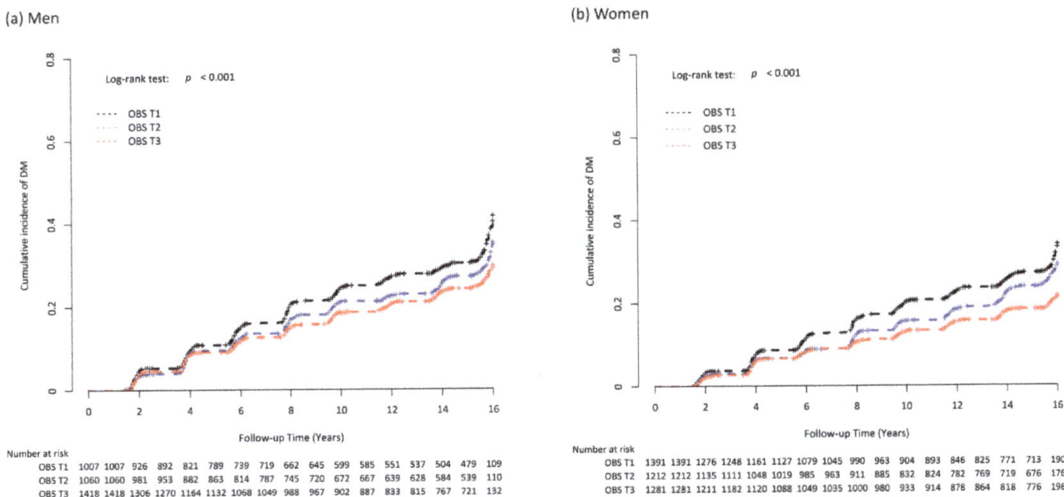

Figure 2. Kaplan–Meier curves for cumulative incidence of type 2 diabetes mellitus according to the sex-specific oxidative balance score tertiles in (**a**) men and (**b**) women.

Table 4. Cox proportional hazard regression analysis presenting the relationship of oxidative balance scores with incident type 2 diabetes mellitus.

Oxidative Balance Score Tertiles	Numbers, *n*	New-Onset Cases, *n*	Follow-Up Period, Person-Year	Incidence Rate Per 1000 Person-Years	Unadjusted	Adjusted
					HR (95% CI)	HR (95% CI)
Men						
Continuous (per 1 increment)					0.94 (0.91–0.96)	0.96 (0.94–0.99)
T1	1007	306	11,130.3	27.49	1 (reference)	1 (reference)
T2	1060	284	12,238.5	23.21	0.85 (0.72–0.99)	0.86 (0.73–1.02)
T3	1418	318	16,218.5	19.61	0.72 (0.62–0.85)	0.83 (0.70–0.99)
p for trend					<0.001	0.035
Women						
Continuous (per 1 increment)					0.91 (0.89–0.94)	0.95 (0.92–0.98)
T1	1391	367	16,206.6	22.65	1 (reference)	1 (reference)
T2	1212	281	14,717.9	19.09	0.84 (0.72–0.98)	0.94 (0.80–1.11)
T3	1281	232	16,025.8	14.48	0.64 (0.54–0.75)	0.78 (0.65–0.94)
p for trend					<0.001	0.010

Adjusted for age, total energy intake, mean blood pressure, education level, household income, plasma fasting glucose, serum insulin, serum total cholesterol, serum triglyceride, and serum C-reactive protein level. Abbreviations: HR, hazard ratio; CI, confidence interval.

4. Discussion

From this prospective study of a large, community-based Korean cohort over 16 years, OBSs were independently and inversely related to incident T2DM even after controlling confounding variables.

In both men and women, the T3 group had 0.83- and 0.78-fold lower HRs for incident T2DM compared with T1 group, respectively. These findings agreed with the results of a previous cross-sectional study, which found that a higher OBS was positively related to better glycemic control in T2DM patients [24]. These data support the hypothesis that a healthy balance of pro- and antioxidant exposure has protection effect against T2DM. To the best of our knowledge, despite the associations between OBS and various health outcomes, including chronic kidney disease [25], hypertension [26], and metabolic syndrome [27], only one cross-sectional study found an association between OBS and glycemic control until the present [24]. A greater OBS, which denotes a predominance of antioxidant exposures over pro-oxidant exposures, has been associated with better glycemic control in Iranian people with T2DM, according to a prior study. [24]. In the prior study, the multivariable-adjusted mean HbA1c and FSG of participants in the highest tertile of OBS were noticeably lower than those in the lowest tertile (for HbA1c: mean difference—0.73 %; for FSG: mean difference—10.2 mg/dL; both $p < 0.050$). However, causal relationships cannot be inferred due to the study's cross-sectional nature. This cross-sectional study was performed on participants who have already been diagnosed with T2DM. Our prospective study is the first approach to evaluate the effect of OBS on the incidence of T2DM in the general population.

In both men and women, the T3 group consumed higher amounts of both antioxidant components (such as omega-3 PUFA, selenium, vitamin C, vitamin E, and beta-carotene) and pro-oxidant components (such as saturated fatty acids, omega-6, and iron) compared with the other groups. This could potentially be attributed to their higher total energy intake. Considering these findings, it is believed that taking into account the OBS is more important than considering the individual components alone. Additionally, one important consideration is that factors like smoking, alcohol consumption, and obesity may have a greater impact on an OBS.

There are several persuasive mechanisms assisting the noted associations with lower risk for T2DM in the current study. Healthy diet patterns emphasizing a high consumption of fruits, vegetables, nuts, and fish are associated with health benefits including improvement of serum glucose and lipid level and weight loss [28]. Fruits, vegetables, nuts, and fish are rich sources of vitamins, minerals, polyphenols, and healthy fats, which have been associated with enhancing insulin sensitivity and reducing inflammation [29].

Physical activity yields a range of favorable effects, including enhancements in serum lipids, peripheral insulin sensitivity, reduction in blood pressure, mitigation of inflammation, and facilitation of weight loss [30]. Smoking can negatively impact pancreatic β-cell function and insulin sensitivity, promote inflammation, and contribute to increased visceral adiposity, in contrast to individuals who do not smoke [31]. Therefore, research groups have provided evidence that adopting a healthy lifestyle, encompassing reduced alcohol consumption, weight control, and increased vegetable intake, can effectively mitigate the risk of developing T2DM among individuals with impaired glucose tolerance and fasting glucose levels [32]. Further recent meta-analysis highlights that combining healthy lifestyles including healthy diet patterns, physical activity, cessation of smoking, and a healthy weight is closely associated with lower risk of T2DM [8].

This study has a few limitations. First, selection bias, as in other prospective studies, could have occurred. The subjects were recruited from 38 health examination centers and hospitals in the Republic of Korea's urban district, and only those willing to perform were enrolled. We could not assess the effects of individual pro- and anti-inflammatory cytokines, including TNF-α, IL-1β, IL-4, IL-6, and IL-10. Second, there is no information in the KoGES on detailed prescriptions for antidiabetic medications. Third, in the KoGES dataset, only the baseline survey data for OBS values were utilized. This was because follow-up information specifically related to diet was unavailable. It is important to note that all variables included in the OBS have the potential to change over time. Therefore, future studies should consider analyzing the impact of changes in OBSs over time on the incidence of T2DM. Forth, each component comprising the oxidative balance score may exert unique effects on the incidence of T2DM. Therefore, it is crucial to employ an analytical approach that incorporates the weights associated with each pro-oxidant and antioxidant component when evaluating their influence on the development of T2DM. Further research is needed to clarify the association between OBS and T2DM. Finally, the indicators included in an OBS can contribute to the development of T2DM not only through oxidative stress effects but also through other mechanisms. For instance, high levels of physical activity have a protective effect against diabetes by improving insulin resistance in the muscles and liver [33]. On the other hand, obesity can contribute to T2DM through altered pancreatic hormone secretion, impaired glucose uptake in skeletal muscles, and hepatic insulin resistance [34]. Therefore, the group with high OBSs may have been influenced by additional mechanisms, beyond oxidative stress, in the occurrence of T2DM. Despite the above limitations, the most notable feature of this prospective study was confirmation of the incidence of T2DM by analyzing FFQ nutritional details on a large scale over 16 years. As a result, it reduces the possibility of recall bias and provides more reliable results than case-control studies. This current study is significant for providing evidence of a positive relationship between OBS and T2DM incidence risk. We anticipate that the present research will help lower the incidence of T2DM by highlighting the importance of an antioxidant-rich diet and drawing public attention to the risk of a pro-inflammatory lifestyle and diet.

5. Conclusions

We found that higher OBS was significantly related to a lower risk of T2DM among community-dwelling middle-aged and older Korean adults. Maintaining an optimal weight, physical activity, a non-smoking lifestyle, and a healthy diet pattern could be effective for lowering T2DM risk.

Author Contributions: Conceptualization, J.-H.L.; methodology, Y.-J.K., H.-M.P., and. J.-H.L.; validation, Y.-J.K., H.-M.P., and. J.-H.L.; formal analysis, Y.-J.K. and J.-H.L.; investigation, H.-M.P. and J.-H.L.; resources, H.-M.P.; data curation, J.-H.L.; writing—original draft preparation, Y.-J.K., H.-M.P., and. J.-H.L.; writing—review and editing, J.-H.L.; visualization, Y.-J.K., H.-M.P., and. J.-H.L.; supervision, J.-H.L.; project administration, Y.-J.K., H.-M.P., and. J.-H.L. All authors have read and agreed to the published version of the manuscript.

Funding: This work was supported by the Korea Institute of Planning and Evaluation for Technology in Food, Agriculture, and Forestry through the High Value-added Food Technology Development Program funded by the Ministry of Agriculture, Food and Rural Affairs (321030051HD030).

Institutional Review Board Statement: This study protocol was approved by the Institutional Review Board of Nowon Eulji Medical Center (Approval number: 2021-09-025).

Informed Consent Statement: Informed consent was obtained from all subjects involved in the study.

Data Availability Statement: The Korean Genome and Epidemiology Study data are available through a procedure described at https://nih.go.kr/ko/main/main.do (accessed on 23 January 2023).

Acknowledgments: Data in this study were obtained from the Korean Genome and Epidemiology Study (4851-302), National Research Institute of Health, Centers for Disease Control and Prevention, Ministry for Health and Welfare, Republic of Korea. Informed consent was obtained from all the participants in the current study.

Conflicts of Interest: No potential conflict of interest was reported by the authors.

References

1. Saeedi, P.; Petersohn, I.; Salpea, P.; Malanda, B.; Karuranga, S.; Unwin, N.; Colagiuri, S.; Guariguata, L.; Motala, A.A.; Ogurtsova, K.; et al. Global and regional diabetes prevalence estimates for 2019 and projections for 2030 and 2045: Results from the international diabetes federation diabetes atlas, 9(th) edition. *Diabetes Res. Clin. Prac.* **2019**, *157*, 107843. [CrossRef] [PubMed]
2. Bae, J.C. Trends of diabetes epidemic in korea. *Diabetes Metab. J.* **2018**, *42*, 377–379. [CrossRef] [PubMed]
3. Oh, S.-H.; Ku, H.; Park, K.S. Prevalence and socioeconomic burden of diabetes mellitus in south korean adults: A population-based study using administrative data. *BMC Public Health* **2021**, *21*, 548. [CrossRef]
4. Cousin, E.; Schmidt, M.I.; Ong, K.L.; Lozano, R.; Afshin, A.; Abushouk, A.I.; Agarwal, G.; Agudelo-Botero, M.; Al-Aly, Z.; Alcalde-Rabanal, J.E.; et al. Burden of diabetes and hyperglycaemia in adults in the americas, 1990–2019: A systematic analysis for the global burden of disease study 2019. *Lancet Diabetes Endocrinol.* **2022**, *10*, 655–667. [CrossRef] [PubMed]
5. Galicia-Garcia, U.; Benito-Vicente, A.; Jebari, S.; Larrea-Sebal, A.; Siddiqi, H.; Uribe, K.B.; Ostolaza, H.; Martin, C. Pathophysiology of type 2 diabetes mellitus. *Int. J. Mol. Sci.* **2020**, *21*, 6275. [CrossRef]
6. Wu, H.; Ballantyne, C.M. Metabolic inflammation and insulin resistance in obesity. *Circ. Res.* **2020**, *126*, 1549–1564. [CrossRef]

7. Rohm, T.V.; Meier, D.T.; Olefsky, J.M.; Donath, M.Y. Inflammation in obesity, diabetes, and related disorders. *Immunity* **2022**, *55*, 31–55. [CrossRef]
8. Khan, T.A.; Field, D.; Chen, V.; Ahmad, S.; Mejia, S.B.; Kahleová, H.; Rahelić, D.; Salas-Salvadó, J.; Leiter, L.A.; Uusitupa, M.; et al. Combination of multiple low-risk lifestyle behaviors and incident type 2 diabetes: A systematic review and dose-response meta-analysis of prospective cohort studies. *Diabetes Care* **2023**, *46*, 643–656. [CrossRef]
9. Goodman, M.; Bostick, R.M.; Dash, C.; Flanders, W.D.; Mandel, J.S. Hypothesis: Oxidative stress score as a combined measure of pro-oxidant and antioxidant exposures. *Ann. Epidemiol.* **2007**, *17*, 394–399. [CrossRef]
10. Hernández-Ruiz, Á.; García-Villanova, B.; Guerra-Hernández, E.; Amiano, P.; Ruiz-Canela, M.; Molina-Montes, E. A review of a priori defined oxidative balance scores relative to their components and impact on health outcomes. *Nutrients* **2019**, *11*, 774. [CrossRef]
11. Lakkur, S.; Goodman, M.; Bostick, R.M.; Citronberg, J.; McClellan, W.; Flanders, W.D.; Judd, S.; Stevens, V.L. Oxidative balance score and risk for incident prostate cancer in a prospective u.S. Cohort study. *Ann. Epidemiol.* **2014**, *24*, 475–478.e474. [CrossRef]
12. Kong, S.Y.; Goodman, M.; Judd, S.; Bostick, R.M.; Flanders, W.D.; McClellan, W. Oxidative balance score as predictor of all-cause, cancer, and noncancer mortality in a biracial us cohort. *Ann. Epidemiol.* **2015**, *25*, 256–262.e251. [CrossRef]
13. Cho, A.R.; Kwon, Y.J.; Lim, H.J.; Lee, H.S.; Kim, S.; Shim, J.Y.; Lee, H.R.; Lee, Y.J. Oxidative balance score and serum γ-glutamyltransferase level among korean adults: A nationwide population-based study. *Eur. J. Nutr.* **2018**, *57*, 1237–1244. [CrossRef]
14. Haggag Mel, S.; Elsanhoty, R.M.; Ramadan, M.F. Impact of dietary oils and fats on lipid peroxidation in liver and blood of albino rats. *Asian Pac. J. Trop Biomed* **2014**, *4*, 52–58. [CrossRef] [PubMed]
15. Pitaraki, E.E. The role of mediterranean diet and its components on the progress of osteoarthritis. *J. Frailty Sarcopenia Falls* **2017**, *2*, 45–52. [CrossRef]
16. Romeu, M.; Aranda, N.; Giralt, M.; Ribot, B.; Nogues, M.R.; Arija, V. Diet, iron biomarkers and oxidative stress in a representative sample of mediterranean population. *Nutr. J.* **2013**, *12*, 102. [CrossRef]
17. Valenzuela, R.; Rincón-Cervera, M.; Echeverría, F.; Barrera, C.; Espinosa, A.; Hernández-Rodas, M.C.; Ortiz, M.; Valenzuela, A.; Videla, L.A. Iron-induced pro-oxidant and pro-lipogenic responses in relation to impaired synthesis and accretion of long-chain polyunsaturated fatty acids in rat hepatic and extrahepatic tissues. *Nutrition* **2018**, *45*, 49–58. [CrossRef] [PubMed]
18. Annor, F.B.; Goodman, M.; Okosun, I.S.; Wilmot, D.W.; Il'yasova, D.; Ndirangu, M.; Lakkur, S. Oxidative stress, oxidative balance score, and hypertension among a racially diverse population. *J. Am. Soc. Hypertens* **2015**, *9*, 592–599. [CrossRef]
19. Kim, Y.; Han, B.G. Cohort profile: The korean genome and epidemiology study (koges) consortium. *Int. J. Epidemiol.* **2017**, *46*, e20. [CrossRef] [PubMed]
20. American Diabetes Association. 2. Classification and diagnosis of diabetes: Standards of medical care in diabetes-2020. *Diabetes Care* **2020**, *43*, S14–S31. [CrossRef]
21. Seo, M.H.; Lee, W.Y.; Kim, S.S.; Kang, J.H.; Kang, J.H.; Kim, K.K.; Kim, B.Y.; Kim, Y.H.; Kim, W.J.; Kim, E.M.; et al. 2018 korean society for the study of obesity guideline for the management of obesity in korea. *J. Obes. Metab. Syndr.* **2019**, *28*, 40. [CrossRef] [PubMed]
22. Oh, J.Y.; Yang, Y.J.; Kim, B.S.; Kang, J.H. Validity and reliability of korean version of international physical activity questionnaire (ipaq) short form. *J. Korean Acad. Fam. Med.* **2007**, *28*, 532–541.
23. Ahn, Y.; Kwon, E.; Shim, J.E.; Park, M.K.; Joo, Y.; Kimm, K.; Park, C.; Kim, D.H. Validation and reproducibility of food frequency questionnaire for korean genome epidemiologic study. *Eur. J. Clin. Nutr.* **2007**, *61*, 1435–1441. [CrossRef] [PubMed]
24. Golmohammadi, M.; Ayremlou, P.; Zarrin, R. Higher oxidative balance score is associated with better glycemic control among iranian adults with type-2 diabetes. *Int. J. Vitam. Nutr. Res.* **2021**, *91*, 31–39. [CrossRef]
25. Son, D.H.; Lee, H.S.; Seol, S.Y.; Lee, Y.J.; Lee, J.H. Association between the oxidative balance score and incident chronic kidney disease in adults. *Antioxidants* **2023**, *12*, 335. [CrossRef]
26. Lee, J.H.; Son, D.H.; Kwon, Y.J. Association between oxidative balance score and new-onset hypertension in adults: A community-based prospective cohort study. *Front. Nutr.* **2022**, *9*, 1066159. [CrossRef]
27. Noruzi, Z.; Jayedi, A.; Farazi, M.; Asgari, E.; Dehghani Firouzabadi, F.; Akbarzadeh, Z.; Djafarian, K.; Shab-Bidar, S. Association of oxidative balance score with the metabolic syndrome in a sample of iranian adults. *Oxid. Med. Cell Longev.* **2021**, *2021*, 5593919. [CrossRef]
28. Chiavaroli, L.; Viguiliouk, E.; Nishi, S.K.; Blanco Mejia, S.; Rahelić, D.; Kahleová, H.; Salas-Salvadó, J.; Kendall, C.W.; Sievenpiper, J.L. Dash dietary pattern and cardiometabolic outcomes: An umbrella review of systematic reviews and meta-analyses. *Nutrients* **2019**, *11*, 338. [CrossRef]
29. Thomas, M.S.; Calle, M.; Fernandez, M.L. Healthy plant-based diets improve dyslipidemias, insulin resistance, and inflammation in metabolic syndrome. A narrative review. *Adv. Nutr.* **2023**, *14*, 44–54. [CrossRef]
30. Chow, L.S.; Gerszten, R.E.; Taylor, J.M.; Pedersen, B.K.; van Praag, H.; Trappe, S.; Febbraio, M.A.; Galis, Z.S.; Gao, Y.; Haus, J.M.; et al. Exerkines in health, resilience and disease. *Nat. Rev. Endocrinol.* **2022**, *18*, 273–289. [CrossRef]
31. Yeh, H.C.; Duncan, B.B.; Schmidt, M.I.; Wang, N.Y.; Brancati, F.L. Smoking, smoking cessation, and risk for type 2 diabetes mellitus: A cohort study. *Ann. Intern. Med.* **2010**, *152*, 10–17. [CrossRef] [PubMed]

32. Davies, M.J.; Aroda, V.R.; Collins, B.S.; Gabbay, R.A.; Green, J.; Maruthur, N.M.; Rosas, S.E.; Del Prato, S.; Mathieu, C.; Mingrone, G.; et al. Management of hyperglycemia in type 2 diabetes, 2022. A consensus report by the american diabetes association (ada) and the european association for the study of diabetes (easd). *Diabetes Care* **2022**, *45*, 2753–2786. [PubMed]
33. Kirwan, J.P.; Sacks, J.; Nieuwoudt, S. The essential role of exercise in the management of type 2 diabetes. *Cleve Clin. J. Med.* **2017**, *84*, S15–S21. [CrossRef]
34. Kahn, S.E.; Hull, R.L.; Utzschneider, K.M. Mechanisms linking obesity to insulin resistance and type 2 diabetes. *Nature* **2006**, *444*, 840–846. [CrossRef] [PubMed]

Disclaimer/Publisher's Note: The statements, opinions and data contained in all publications are solely those of the individual author(s) and contributor(s) and not of MDPI and/or the editor(s). MDPI and/or the editor(s) disclaim responsibility for any injury to people or property resulting from any ideas, methods, instructions or products referred to in the content.

Review

Potential Benefits of Selenium Supplementation in Reducing Insulin Resistance in Patients with Cardiometabolic Diseases: A Systematic Review and Meta-Analysis

Jiahui Ouyang [1,†], Yajie Cai [1,†], Yewen Song [1], Zhuye Gao [1,2], Ruina Bai [1,2,*] and Anlu Wang [1,2,*]

1 Xiyuan Hospital, China Academy of Chinese Medical Sciences, Beijing 100091, China
2 National Clinical Research Center for Chinese Medicine Cardiology, Xiyuan Hospital, China Academy of Chinese Medical Sciences, Beijing 100091, China
* Correspondence: brntcl@126.com (R.B.); wanganlu@bucm.edu.cn (A.W.)
† These authors contributed equally to this work.

Abstract: Background: Selenium is a trace element that has been reported to be effective in regulating glucose and lipid metabolism. However, there is conflicting evidence from different clinical trials of selenium supplementation in treating cardiometabolic diseases (CMDs). Objective: This meta-analysis aimed to identify the effects of selenium supplementation on insulin resistance, glucose homeostasis, and lipid profiles in patients with CMDs. Methods: Randomized controlled trials (RCTs) of selenium supplementation for treating CMDs were screened in five electronic databases. Insulin levels, homeostatic model assessment of insulin resistance (HOMA-IR), fasting plasma glucose (FPG), and glycosylated hemoglobin A1C (HbA1c) were defined as the primary outcome markers, and lipid profiles were considered the secondary outcome markers. Results: Ten studies involving 526 participants were included in the meta-analysis. The results suggested that selenium supplementation significantly reduced serum insulin levels (standardized men difference [SMD]: −0.53; 95% confidence interval [CI] [−0.84, −0.21], $p = 0.001$, $I^2 = 68\%$) and HOMA-IR (SMD: −0.50, 95% CI [−0.86, −0.14], $p = 0.006$, $I^2 = 75\%$) and increased high-density lipoprotein cholesterol (HDL-C) levels (SMD: 0.97; 95% CI [0.26, 1.68], $p = 0.007$, $I^2 = 92\%$), but had no significant effect on FPG, total cholesterol (TC), triglycerides (TG), low-density lipoprotein cholesterol (LDL-C), and very low-density lipoprotein cholesterol (VLDL-C). Conclusion: Current evidence supports the beneficial effects of selenium supplementation on reducing insulin levels, HOMA-IR, and increasing HDL-C levels. Selenium supplementation may be an effective strategy for reducing insulin resistance in patients with CMDs. However, more high-quality clinical studies are needed to improve the certainty of our estimates.

Keywords: selenium; cardiometabolic disease; insulin resistance; diabetes mellitus; cardiovascular disease; systematic review; meta-analysis

Citation: Ouyang, J.; Cai, Y.; Song, Y.; Gao, Z.; Bai, R.; Wang, A. Potential Benefits of Selenium Supplementation in Reducing Insulin Resistance in Patients with Cardiometabolic Diseases: A Systematic Review and Meta-Analysis. *Nutrients* **2022**, *14*, 4933. https://doi.org/10.3390/nu14224933

Academic Editor: Silvia V. Conde

Received: 15 October 2022
Accepted: 18 November 2022
Published: 21 November 2022

Publisher's Note: MDPI stays neutral with regard to jurisdictional claims in published maps and institutional affiliations.

Copyright: © 2022 by the authors. Licensee MDPI, Basel, Switzerland. This article is an open access article distributed under the terms and conditions of the Creative Commons Attribution (CC BY) license (https://creativecommons.org/licenses/by/4.0/).

1. Introduction

Cardiometabolic diseases (CMDs) begin with clinically high-risk states ranging from insulin resistance (IR) to prediabetes states (e.g., obesity) and metabolic syndrome (MS), which then progress to type 2 diabetes mellitus (T2DM) and cardiovascular disease (CVD) [1]. Most individuals diagnosed with CMDs also experience additional cardiometabolic risks, including obesity, dyslipidemia, hyperinsulinemia, and thrombosis [2]. Unhealthy diets and accelerated population aging have resulted in an annual increase in the prevalence and mortality of CMDs, which places a significant economic burden on the healthcare systems of various countries [3,4]. Studies have shown that these metabolic disorders may be influenced by nutrition. Individual nutrient and dietary supplements composed of sufficient nutrients may be associated with reduced cardiometabolic risks, suggesting that nutritional intervention measures are effective in managing these diseases [5–8].

Selenium is a micronutrient that is vital for human health, and its content in soil varies across different regions. In many countries, including China, selenium-containing soil is poor, and people are prone to selenium deficiency [9,10]. Selenium can exist in nature as inorganic forms (e.g., selenate and selenite) and organic forms (e.g., selenocysteine [Sec]) [11]. Selenium is highly absorbed and distributed throughout the body; in particular, organic selenium is more stable and bioavailable than inorganic selenium [11]. Sec is the main form of selenium in cells, and Sec-containing protein, namely selenoprotein (e.g., selenoprotein P, glutathione peroxidases [GPxs], and thioredoxin reductase [TrxRs]), is mainly responsible for the biological role of selenium in the human body [11–14]. Selenium is incorporated into selenoproteins, which have broad pleiotropic functions, such as antioxidant and anti-inflammatory properties [15]. Oxidative stress, which is an imbalance between antioxidant defense and prooxidant substances (e.g., reactive oxygen species [ROS] and reactive nitrogen species [RNS]), causes oxidative damage through various mechanisms, including lipid peroxidative damage, DNA damage, and protein oxidation [16,17]. Lately, increasing evidence has shown that the progression of insulin resistance, the pathogenesis of T2DM, and its microvascular ailments and macrovascular complications are significantly regulated by oxidative stress [18,19]. Higher oxidative stress is directly related to the emergence of CMDs [20,21]. Selenium is a well-known antioxidant; in particular, the selenoproteins GPxs and TrxRs are involved in antioxidant defenses and protection against oxidative damage [22]. Supplementing selenium could significantly reduce ROS, increase superoxide dismutase and GPxs activity, and reduce inflammatory cytokine content [23]. A meta-analysis of 13 trials found that selenium supplementation alleviated oxidative stress by raising the total antioxidant capacity and GPxs levels and lowering serum malonaldehyde [17]. It has been proposed that selenium is a hormetic chemical, a substance with a biphasic dose-response that is poisonous at high levels but beneficial at low concentrations [22]. Supra-nutritional levels of selenium produce ROS, which then disturb the redox states of cells [24], increase oxidative stress, and damage tissues and organs [25]. Therefore, maintaining an optimal selenium status is crucial to maintaining redox equilibrium.

Numerous studies have highlighted the significance of selenium and selenoproteins in the prevention and treatment of chronic metabolic diseases such as MS, T2DM, and CVD [26–28]. Huang et al. [29] reported that low selenium levels were related to an elevated risk of metabolic disorders, poor prognosis, and mortality. Interestingly, Kamali et al. [30] observed that selenium supplementation significantly improved glucose metabolism by decreasing fasting plasma glucose (FPG), insulin, and homeostatic model assessment of insulin resistance (HOMA-IR), and also increased high-density lipoprotein-cholesterol (HDL-C) levels, but did not affect other lipid profiles. However, selenium status has been reported to be positively associated with markers of insulin resistance and lipid profiles by Cardoso et al. [31] and Ju et al. [32]. On the other hand, a previous meta-analysis reported that selenium supplementation significantly alleviated oxidative stress and inflammation, but did not improve the blood lipid status [33]. To the best of our knowledge, the exact role of selenium in glycolipid metabolism in patients with CMDs remains undetermined. Therefore, to address these issues, we analyzed the impacts of selenium supplementation on glucose and lipid metabolism in CMDs, with the aim of verifying whether selenium supplementation could be a complementary treatment strategy for CMDs.

2. Methods

This meta-analysis strictly followed the Preferred Reporting Items for Systematic Reviews and Meta-Analyses (PRISMA) guidelines [34]. The study protocol has been registered and published in PROSPERO with ID: CRD42022353393.

2.1. Search Strategy

Searches of the literature for this meta-analysis were conducted using PubMed, Cochrane Library, Embase, Scopus, and Web of Science databases up to 31 July 2022.

The key search terms for searching the databases included the following: selenium, selenite, selenate, trace element, cardiometabolic disease, diabetes mellitus, T2DM, coronary heart disease, heart failure, hypertension, hyperlipidemia, metabolic syndrome, stroke, obesity, randomized controlled trial, RCT, random*. In some cases, we may have added or changed the retrieved keywords depending on the characteristics of the databases (Supplementary Table S1). Moreover, we manually checked the reference lists of the eligible articles to identify extra pertinent research. Two reviewers (J.O. and Y.C.) conducted the literature search independently, and any discrepancies were resolved by consensus.

2.2. Study Selection

Two authors (J.O. and Y.C.) individually filtered all eligible studies using strict inclusion and exclusion criteria. Any differences in opinion were settled through consensus or discussion with Drs. Bai and Wang. The reasons for the exclusion of studies in each phase were recorded. Eligible studies were required to meet the following inclusion criteria according to PICOS: (1) Types of population (P): patients with diseases related to CMDs, such as diabetes mellitus, coronary heart disease, heart failure, hypertension, stroke, metabolic syndrome, and obesity. (2) Types of interventions (I): the experimental group received selenium supplementation but the control group did not. Selenium supplementation in all forms, including inorganic, organic, synthetic, and selenium-enriched yeast, was considered. The treatment dose and period were not limited. (3) Types of comparison (C): the control group received placebo or conventional treatment. (4) Types of outcomes (O): primary outcomes: insulin levels, HOMA-IR, FPG, and glycosylated hemoglobin A1C (HbA1c); secondary outcomes: lipid profiles, including total cholesterol (TC), triglycerides (TG), low-density lipoprotein cholesterol (LDL-C), very low-density lipoprotein cholesterol (VLDL-C), and HDL-C. (5) Types of study design (S): randomized controlled trial (RCT) only. The exclusion criteria were as follows: (1) Repeat published studies; (2) conference abstracts; and (3) in vitro and animal studies.

2.3. Data Collection Process

Two reviewers (J.O. and Y.C.) independently collected the following data from the included studies using standardized forms: (1) the characteristics of selected articles, such as author(s), journal of publication, publication year, study design, study location, registration, or not, the number of participants, interventions, treatment period; (2) characteristics of participants, such as disease type, mean age, gender; and (3) clinical outcomes.

2.4. Risk of Bias Assessment

We assessed the risk of bias of the included studies according to the Cochrane Collaboration Risk of Bias tool. The assessed domains included the following: methods of random sequence generation, allocation concealment, blinding of participants and personnel, blinding of outcome assessment, incomplete outcome data, selective reporting, and other bias [35]. Each study was classified as low, unclear, or high risk based on these domains.

2.5. Data Synthesis and Statistical Analysis

The effect of selenium supplementation on relevant outcomes was assessed as the changes (mean $\pm$ standard deviation [SD]) before and after treatment in the experimental and the control groups. If the mean values of the changes before and after treatment were unreported, they were calculated by subtracting the mean at the baseline from the mean at the end of the follow-up. When the SDs of the changes before and after treatment were not reported, they were computed according to the number of patients, standard errors, 95% confidence interval (CI), interquartile ranges, or *p*-values. If the missing SDs were still unavailable, they were calculated using the correlation formula, and the correlation coefficient was cautiously assumed to be 0.5 [36,37]. For studies with multiple intervention groups, we combined relevant groups into a single treatment group. All related calculation formulas were referred to the Cochrane Handbook for Systematic Reviews of Intervention [38].

Data were evaluated using Review Manager version 5.3 and STATA version 17.0 for a more comprehensive assessment of outcomes. The heterogeneity between studies was assessed using Cochrane's Q test and was quantified by the I^2 test. Heterogeneity was rated as low, moderate, or high when the value of I^2 was <50%, 50–75%, or >75%, respectively [39]. When the heterogeneity was low ($I^2 < 50\%$), data were pooled by applying the fixed-effects model; otherwise, the random-effects model was applied [40]. Effect sizes are presented as the standardized mean difference (SMD) with 95% CI. If sufficient studies ($\geq$10) were included, funnel plots and Egger's test were applied to determine whether there was publication bias. A p-value of <0.05 was considered statistically significant.

2.6. Analysis of Subgroups or Subsets

In cases where significant heterogeneity was noted among studies, sensitivity analysis or subgroup analyses were performed to identify its possible sources. Sensitivity analysis was performed by removing each study sequentially to evaluate the influence of each study on the overall effect size. Subgroup analysis was conducted according to the type of disease of the participants.

3. Results

3.1. Literature Selection

We retrieved 4688 studies from five electronic databases. Two studies were manually retrieved. Then, 2530 studies were retained after excluding 2160 duplicates, and a further 2503 studies were eliminated after reading the title and abstract, leaving 27 studies that met the screening criteria for full-text evaluation. Finally, 10 RCTs [30,41–49] were included in this meta-analysis. The screening process is depicted in Figure 1.

3.2. Study Characteristics

All 10 studies included in this meta-analysis were randomized, double-blind, placebo-controlled trials, with 526 participants, including 272 in the selenium group (experimental group) and 254 in the control group. The treatment period ranged from 4 to 24 weeks. All 10 studies were conducted in Iran. Except for Faghihi 2014 [48], the remaining nine studies have completed clinical trial registration. Faghihi 2014 [48] reported participants' selenium concentration as deficient state at baseline, and the remaining studies did not report participants' selenium status. In the included studies, the forms of selenium supplementation were mainly selenium yeast and sodium selenite, but three studies did not mention the form of selenium supplementation. Five studies [37,41,43,45,48] recruited patients with diabetes mellitus or complications of diabetes mellitus (e.g., diabetic nephropathy), three studies [30,42,44] recruited patients with cardiovascular disease, one study [46] recruited patients with diabetes mellitus combined with coronary heart disease, and one study [49] recruited obese patients (Table 1).

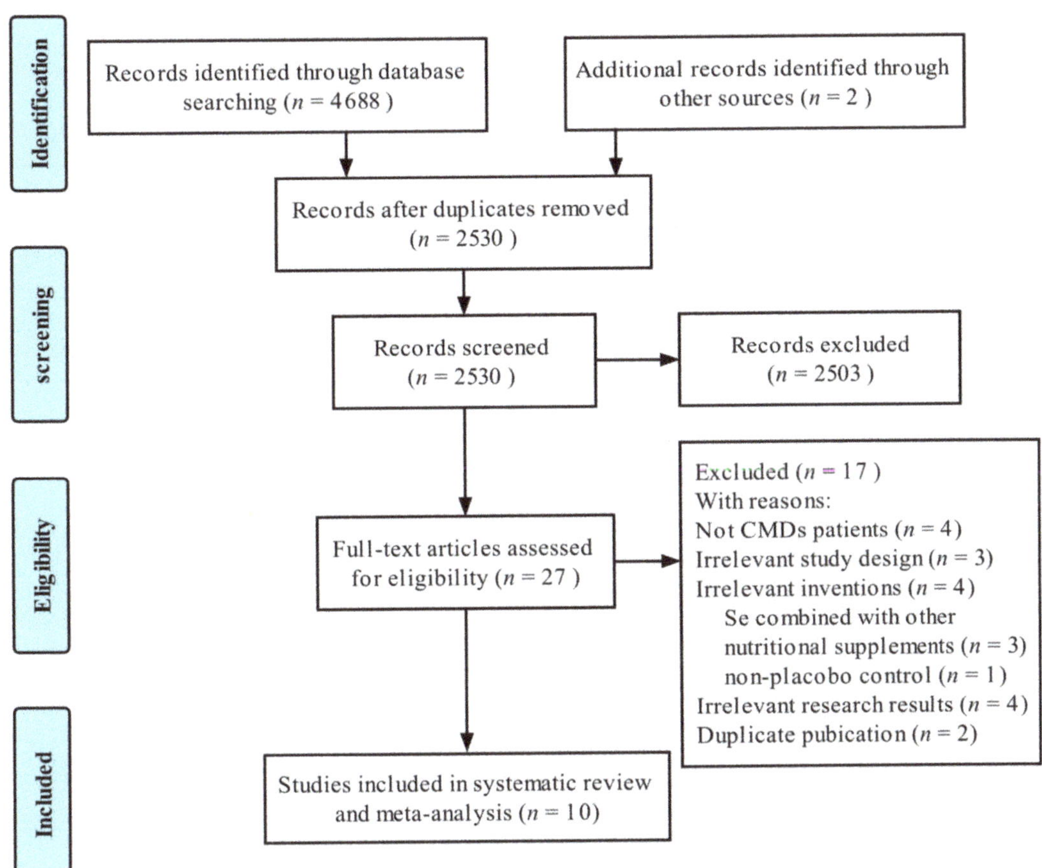

Figure 1. PRISMA flow chart for selection and screening of the studies.

Table 1. Characteristics of the included studies.

Study	Study Location	Study Design	Groups	No. of Participants	Mean Age	Gender (M/F)	Intervention	Treatment Period	Type of Disease
Salimian 2022 [41]	Iran	Single-center double-blinded RCT	I	26	57.6 ± 11.5	NR	selenized yeast 200 µg/day	24 weeks	Diabetic nephropathy
			C	27	61.5 ± 9.8	NR	placebo		
Ghazi 2021 [42]	Iran	Single-center double-blinded RCT	I_1	16	59.06 ± 8.55	12/4	selenium-enriched yeast 200 µg/day	8 weeks	Atherosclerosis
			I_2	16	58.62 ± 9.68	16/0	sodium selenite 200 µg/day		
			C	17	53.58 ± 13.75	15/2	Placebo		
Najib 2020 [43]	Iran	Multi-center double-blinded RCT	I	26	29.19 ± 6.16	0/26	selenium supplements 100 µg/day	12 weeks	Gestational diabetes mellitus
			C	28	31.0 ± 4.43	0/28	Placebo		
Kamali 2019 [30]	Iran	Single-center double-blinded RCT	I	17	62.6 ± 11.6	NR	selenium yeast 200 µg/day	4 weeks	Coronary heart disease
			C	16	61.2 ± 4.6	NR	Placebo		
Raygan 2018 [44]	Iran	Single-center double-blinded RCT	I	26	70.7 ± 10.3	8/18	selenium yeast 200 µg/day	12 weeks	Congestive heart failure
			C	27	68.5 ± 7.7	8/19	Placebo		
Bahmani 2016 [45]	Iran	Single-center double-blinded RCT	I	30	63.1 ± 12.6	15/15	selenium supplements 200 µg/day	12 weeks	Diabetic nephropathy
			C	30	61.4 ± 9.3	15/15	Placebo		
Farrokhian 2016 [46]	Iran	Single-center double-blinded RCT	I	30	NR	10/20	selenium yeast 200µg/day	8 weeks	Type 2 diabetes mellitus and coronary heart disease
			C	30	NR	10/20	Placebo		
Asemi 2015 [47]	Iran	Single-center double-blinded RCT	I	35	27.6 ± 5.3	0/35	selenium supplements 200 µg/day	6 weeks	Gestational diabetes mellitus
			C	35	29.6 ± 3.6	0/35	Placebo		

Table 1. *Cont.*

Study	Study Location	Study Design	Groups	No. of Participants	Mean Age	Gender (M/F)	Intervention	Treatment Period	Type of Disease
Faghihi 2014 [48]	Iran	Single-center double-blinded RCT	I	33	53.54 ± 7.52	16/17	sodium selenite 200 µg/day	3 months	Type 2 diabetes mellitus
			C	27	55.76 ± 7.77	18/9	Placebo		
Alizadeh 2012 [49]	Iran	Single-center double-blinded RCT	I	17	36.6 ± 8.6	0/17	selenium-enriched yeast 200 µg/day	6 weeks	Obesity
			C	17	36.7 ± 8.3	0/17	Placebo		

M: Male, F: Female, I: Intervention, C: Control, NR: Not reported, RCT: Randomized controlled trial.

3.3. Risk of Bias Assessments

All except two studies [46,49] were rated as having a low risk of selection bias for adopting appropriate random sequence generation and allocation concealment methods. Farrokhian 2016 [46] and Alizadeh 2012 [49] were assessed as having an unclear risk of selection bias because they reported random sequence generation methods, but did not report allocation concealment methods. All of the 10 studies were rated as carrying a low risk of performance bias and concealment bias due to complete reporting of blinding implementation. Eight studies [30,41,42,44–47,49] were rated as having a low risk of attrition bias, but Najib 2020 [43] and Faghihi 2014 [48] were rated as high risk due to unbalanced and unexplained loss at follow-up. Farrokhian 2016 [46] and Alizadeh 2012 [49] were rated as high risk of reporting bias because the primary outcomes were changed after the protocol registration. Due to the lack of a registered protocol [48] or the inability to report several secondary outcomes [41–43,45], five studies were rated as unknown risks of report bias. Farrokhian 2016 [45] was rated a high risk of other bias because of the inconsistency in the types of hypoglycemic drugs taken between the selenium and control groups, which may have affected the effect evaluation (Figures 2 and 3).

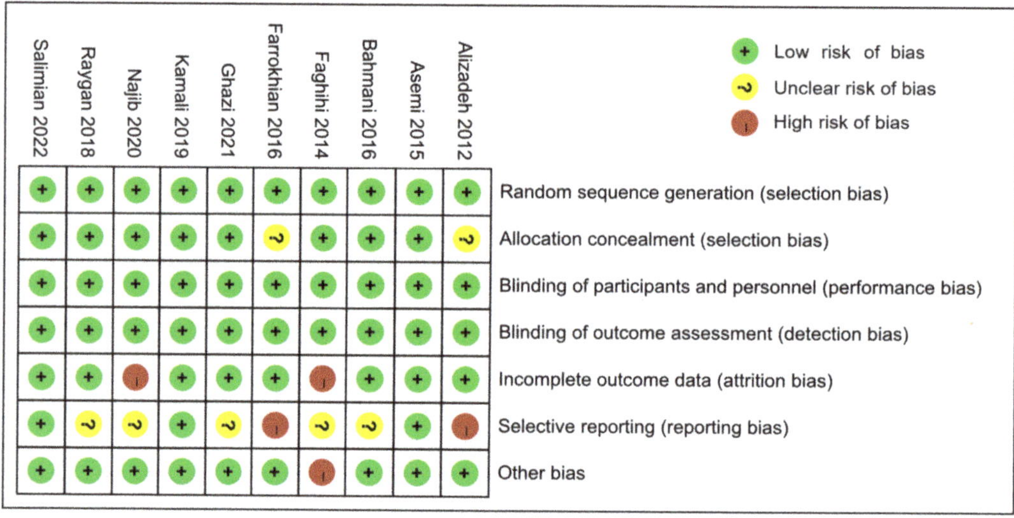

Figure 2. Summary of the risks of bias.

3.4. Meta-Analysis

3.4.1. Primary Outcomes

All 10 studies with 526 participants evaluated the effects of serum insulin levels and HOMA-IR. The heterogeneity of serum insulin levels and HOMA-IR were moderate (I^2 = 68%, I^2 = 75%). Pooled results obtained by employing a random-effects model demonstrated that selenium supplementation remarkably lowered serum insulin levels (SMD: −0.53, 95% CI [−0.84, −0.21], p = 0.001) and decreased HOMA-IR (SMD: −0.50, 95% CI [−0.86, −0.14], p = 0.006) (Figures 4 and 5). To resolve heterogeneity, sensitivity analysis was conducted by excluding the studies one by one. The pooled results were broadly consistent with the above analysis (Supplementary Figures S1 and S2), and the heterogeneity was largely affected by Faghihi 2014 [48], which was excluded. Nine studies [30,41–47,49] remained after the exclusion, with no heterogeneity in the pooled results (all I^2 = 0%), indicating that Faghihi 2014 [48] was a major factor in the source of heterogeneity of insulin levels and HOMA-IR. This may be due to a baseline difference in the hypoglycemic drugs taken between the selenium and control groups in Faghihi's study. An analysis was then conducted with the fixed-effects model, and the result confirmed the previous observation that supplementing with selenium was associated with lower serum insulin levels (SMD: −0.67, 95% CI [−0.86, −0.48], $p < 0.0001$) and HOMA-IR (SMD: −0.67, 95% CI [−0.86, −0.48], $p < 0.0001$) (Figures 6 and 7).

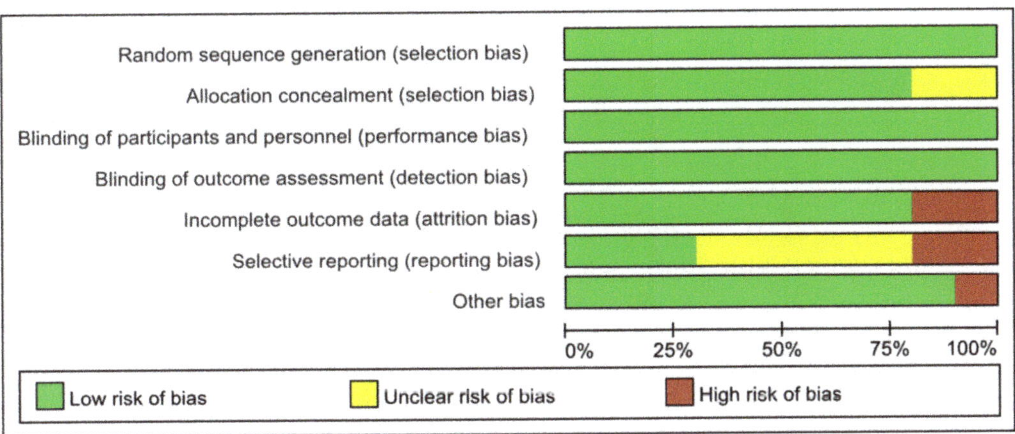

Figure 3. Risks of bias graph expressed as percentages.

Study or Subgroup	Selenium Mean	Selenium SD	Selenium Total	Control Mean	Control SD	Control Total	Weight	Std. Mean Difference IV, Random, 95% CI
Salimian 2022	−0.9	2	26	−0.1	0.9	27	10.3%	−0.51 [−1.06, 0.04]
Ghazi 2021	−5.44	6.77	32	−1.43	0.04	17	9.6%	−0.72 [−1.32, −0.11]
Najib 2020	−1.13	3.81	26	−0.01	2.58	28	10.4%	−0.34 [−0.88, 0.20]
Kamali 2019	−0.9	2.07	17	0.3	2.21	16	8.6%	−0.55 [−1.24, 0.15]
Raygan 2018	−18.41	27.53	26	13.73	23.63	27	9.8%	−1.24 [−1.83, −0.64]
Bahmani 2016	−3.1	4.6	30	0.5	6.2	30	10.6%	−0.65 [−1.17, −0.13]
Farrokhian 2016	−2.2	4.6	30	3.6	8.4	30	10.5%	−0.85 [−1.38, −0.32]
Asemi 2015	−1.98	11.25	35	5.26	9.33	35	11.0%	−0.69 [−1.18, −0.21]
Faghihi 2014	−1.19	5.64	33	−5.29	6.93	27	10.6%	0.65 [0.12, 1.17]
Alizadeh 2012	−3.4	10.0	17	0.2	5.55	17	8.8%	−0.43 [−1.12, 0.25]
Total (95% CI)			272			254	100.0%	−0.53[−0.84, −0.21]

Heterogeneity: $Tau^2 = 0.17$; $Chi^2 = 27.88$, $df = 9$ ($p = 0.0010$); $I^2 = 68\%$
Test for overall effect: $Z = 3.29$ ($p = 0.0010$)

Std. Mean Difference IV, Random, 95% CI: −2, −1, 0, 1, 2; Favours [selenium] Favours [control]

Figure 4. Forest plot of insulin levels [30,41–49].

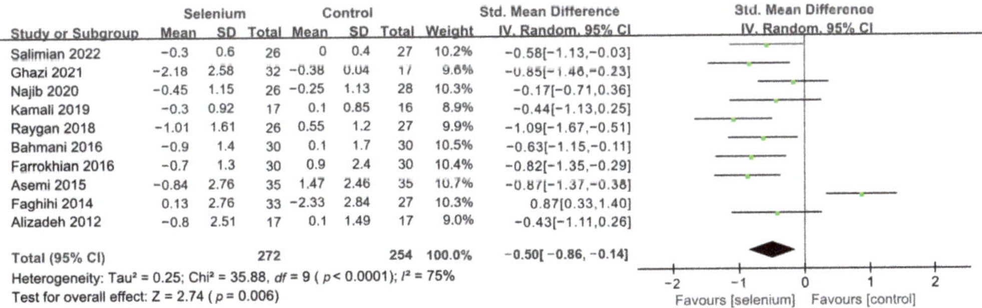

Figure 5. Forest plot of HOMA-IR [30,41–49].

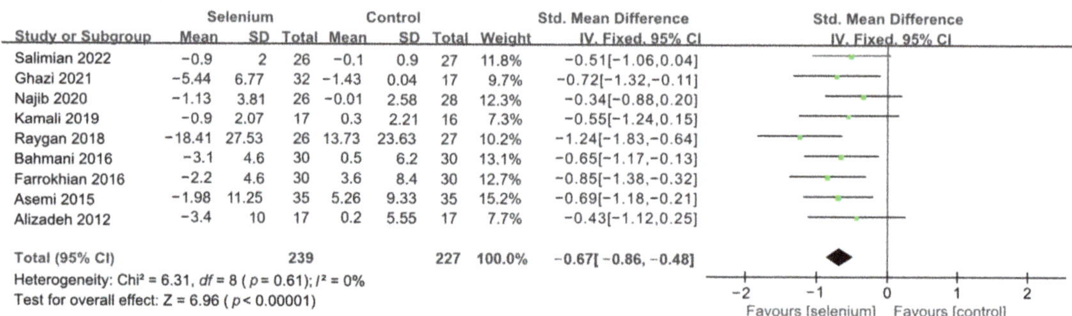

Figure 6. Forest plot of insulin levels after excluding Faghihi 2014 [30,41–47,49].

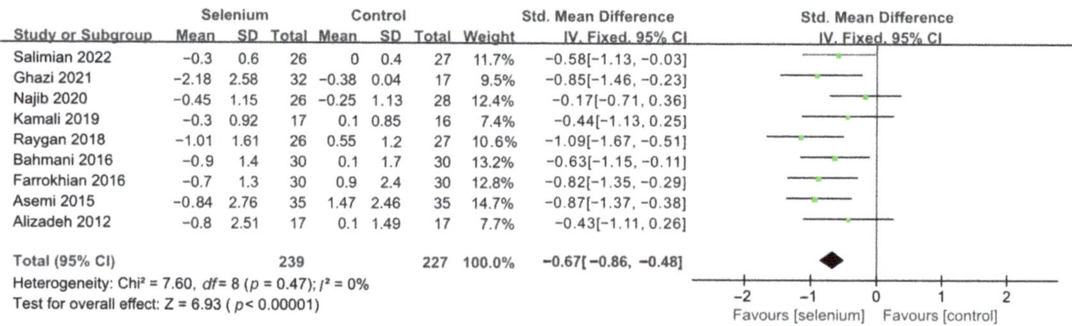

Figure 7. Forest plot of HOMA-IR after excluding Faghihi 2014 [30,41–47,49].

The effect of selenium supplementation on FPG was assessed in 492 participants through nine studies [30,41–48]. The heterogeneity between studies was high (I^2 = 91%). Pooled analysis from the random-effects model indicated that the selenium group and the control group had similar effects on FPG (SMD: 0.06, 95% CI [−0.56, 0.68], p = 0.86) (Supplementary Figure S3). To resolve heterogeneity, sensitivity analysis was conducted by excluding the studies one by one. The results showed that although the pooled results were stable, the heterogeneity was not resolved (Supplementary Figure S4). Then, subgroup analysis was conducted based on the underlying diseases of the participants. As shown in Figure 8, the FPG levels in patients with cardiovascular disease were significantly lower in the selenium group than in the control group (SMD: −0.42, 95% CI: [−0.77, −0.07], p = 0.02), with no heterogeneity (I^2 = 0%). However, there was no statistical difference in terms of FPG between the selenium and control groups in the other two subgroups (Figure 8).

Only two studies [43,48], including 114 participants, assessed the effect of selenium supplementation on HbA1c. The heterogeneity between studies was high (I^2 = 85%). Thus, we did not perform a meta-analysis of HbA1c. Both studies reported reductions in HbA1c after treatment in both the selenium and control groups, in which Najib 2020 [43] reported a more significant reduction in HbA1c in the selenium group compared to the control group.

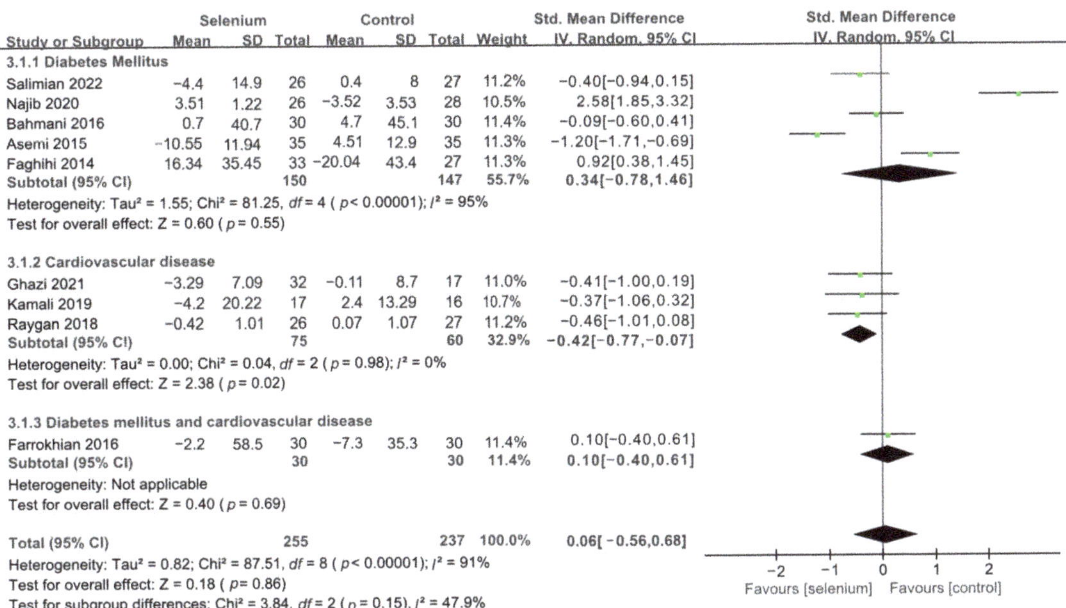

Figure 8. Subgroup analysis of FPG [30,41–48].

3.4.2. Secondary Outcomes

Nine studies [30,41,42,44–49] with 492 patients evaluated the effects of TC, TG, and LDL-C, and five studies [30,41,44–46] with 259 patients evaluated the effects of VLDL-C. The heterogeneity of TC, TG, and VLDL-C was insignificant (all I^2 = 0). Unfortunately, the pooled results from the fixed-effects model demonstrated that selenium supplements did not significantly lower TC, TG, and VLDL-C in patients with CMDs (SMD: −0.07, 95% CI [−0.25, 0.12], p = 0.48, SMD: −0.12, 95% CI [−0.30, 0.06], p = 0.20, and SMD: −0.08, 95% CI [−0.33, 0.16], p = 0.51, respectively) (Figures 9–11). The heterogeneity of LDL-C was high (I^2 = 79%). Pooled results from the random-effects model demonstrated no significant difference in LDL-C between the two groups (SMD: −0.35, 95% CI [−0.76, 0.06], p = 0.10) (Figure 12).

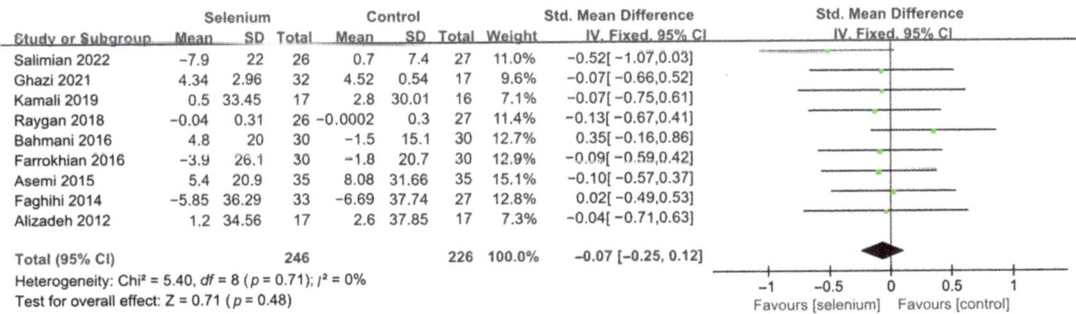

Figure 9. Forest plot of TC [30,41,42,44–49].

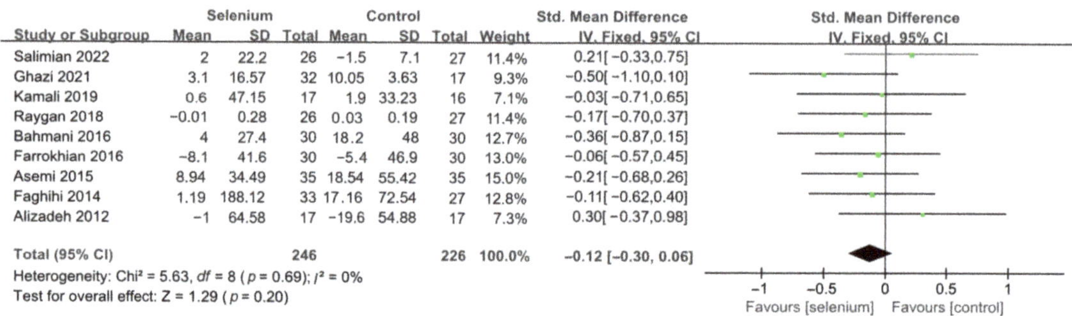

Figure 10. Forest plot of TG [30,41,42,44–49].

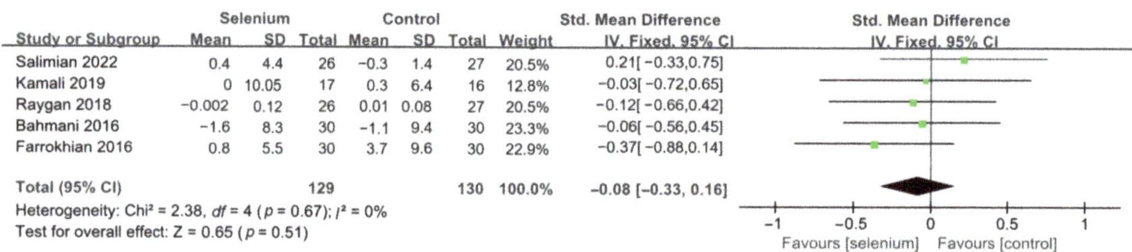

Figure 11. Forest plot of VLDL-C [30,41,44–46].

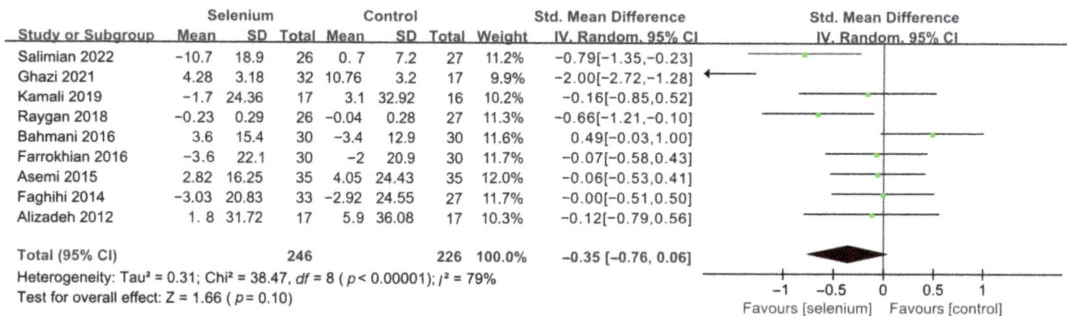

Figure 12. Forest plot of LDL-C [30,41,42,44–49].

A total of nine studies [30,41,42,44–49], including 492 patients, evaluated the effects of HDL-C. The pooled results from the random-effects model indicated that selenium supplementation remarkably increased HDL-C levels (SMD: 0.97, 95% CI [0.26, 1.68], $p = 0.007$), with high heterogeneity across studies ($I^2 = 92\%$) (Supplementary Figure S5). To resolve heterogeneity, a sensitivity analysis was conducted by excluding each study separately. The results showed that the pooled results were broadly consistent with the above analysis (Supplementary Figure S6), and the heterogeneity was largely affected by Ghazi 2021 [42]. Eight studies [30,41,44–49] remained after excluding Ghazi 2021 [42], and pooled results showed that the heterogeneity between studies was decreased ($I^2 = 58\%$) (Supplementary Figure S7). The participants of Ghazi 2021 [42] were patients with atherosclerosis, and dyslipidemia is closely related to atherosclerosis. Therefore, considering that the source of heterogeneity may be related to the participants' underlying disease, the included studies were divided into subgroups based on the participants' disease type. As shown in Figure 13, the HDL-C levels were significantly increased in the diabetes mellitus subgroup and cardiovascular disease subgroup (SMD: 0.50, 95% CI [0.10, 0.91], $p = 0.02$ and SMD:

4.22, 95% CI [1.06, 7.37], $p = 0.009$), with significant heterogeneity among studies ($I^2 = 59\%$ and $I^2 = 97\%$). However, there was no statistical difference between the selenium and control groups in the other two subgroups (Figure 13).

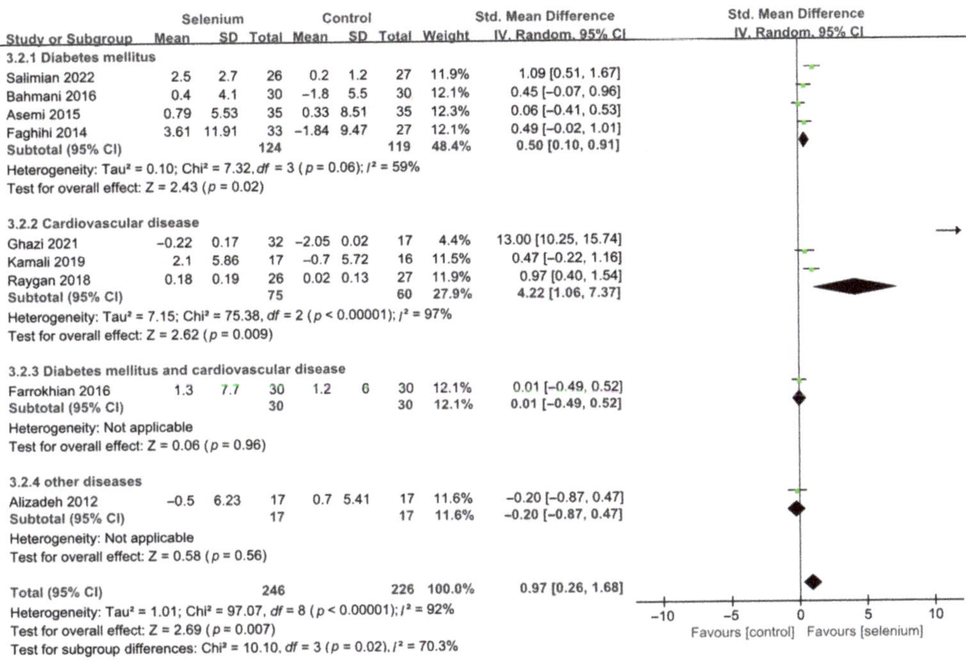

Figure 13. Subgroup analysis of HDL-C [30,41,42,44–49].

3.5. Publication Bias

Funnel plots of insulin levels and HOMA-IR were drawn with Review Manager version 5.3, and Egger's test was conducted to quantify the publication bias with Stata version 17.0. The two funnel plots were substantially symmetrical (Figure 14). The results of Egger's test for insulin levels and HOMA-IR were $p = 0.678$ and $p = 0.908$, respectively, indicating that there was no publication bias.

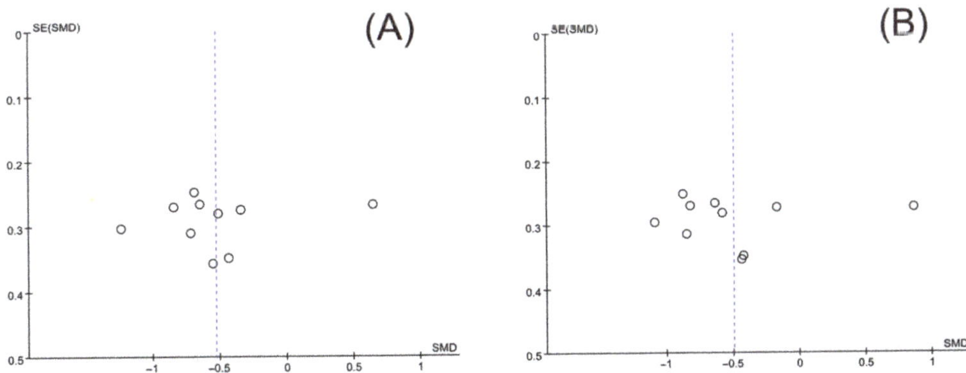

Figure 14. Funnel plot of insulin levels and HOMA-IR. (**A**) Insulin levels, (**B**) HOMA-IR.

4. Discussion

Over the last decade, increasing attention has been paid to the selenium status in patients with various cardiometabolic diseases and the association between selenium status and mortality risk of various cardiometabolic diseases. Several meta-analyses have indicated that individuals with cardiometabolic diseases tend to have lower selenium levels than healthy individuals [50–52], and that selenium levels are negatively associated with mortality among patients with MS, T2DM, and CVD [53–55]. Alterations in the metabolism of glucose and lipids characterize metabolic disorders [56]. Insulin has regulatory effects on glucose, which are mainly classified into two aspects: promoting glucose absorption into liver cells, muscle cells, and adipose tissue, and inhibiting glycogenolysis and gluconeogenesis in the liver [57]. Insulin resistance is a common pathophysiological status in which individuals have decreased insulin sensitivity and impaired glucose regulation by insulin, resulting in glucose intolerance [58]. It is well-established that insulin resistance underpins many chronic metabolic diseases [59], and cardiometabolic risks such as obesity and dyslipidemia can exacerbate insulin resistance and impel the progression of CMDs [1]. Studies have shown that the micronutrient selenium can regulate the human body's insulin sensitivity, and selenium in the form of selenate is known to act as an insulin mimetic with a role in maintaining blood sugar homeostasis [60,61]. Expression of selenoprotein *P* plays a crucial role in pancreatic β-cell function by preventing ferroptosis and thus maintaining glutathione peroxidase 4 (Gpx4) and cell viability, and also by inhibiting stress-induced degradation of nascent granules, a novel regulatory pathway for insulin [62], thus maintaining cellular proinsulin levels [63]. Furthermore, a meta-analysis performed by Tabrizi et al. [64] supported the positive effect of selenium supplements on lipid profiles. Of note, that selenium supplementation makes the most sense when there is deficit of selenium [14,65].

In this meta-analysis, we examined the impact of selenium supplementation on the markers of insulin resistance, glucose, as well as blood lipid profiles in patients with CMDs. The level of glucose in the blood is one of the most significant homeostatic indicators and is strictly regulated [66]. The pathways and regulation of glucose metabolism include glycolysis/glycogenolysis, gluconeogenesis, pentose phosphate pathway (PPP), insulin signaling pathway, and many others [67], some of which can be regulated by insulin [57]. HOMA-IR is often used in investigating and quantifying insulin resistance because of the simplicity of the underlying mathematical model (HOMA-IR = fasting glucose [mg] × fasting insulin [mu/L]/22.5) [68]. Furthermore, HbA1c is an essential marker of long-term glycemic control that reflects a cumulative glycemic level over the past two to three months [69]. Therefore, insulin levels, HOMA-IR, FPG and HbA1c were used to evaluate the selenium supplementation on glycemic control. In this study, comprehensive pooled results from 10 RCTs involving 526 patients supported the favorable effects of selenium supplementation in decreasing serum insulin levels and HOMA-IR. Moreover, selenium supplementation may increase HDL-C levels, but the effectiveness of selenium supplementation on FPG, TC, TG, LDL-C, and VLDL-C levels was unclear. The current results suggest that selenium supplementation may be an effective treatment for reducing insulin resistance.

The findings reveal that selenium supplementation could reduce insulin levels and HOMA-IR in patients with CMDs, but the effect on FPG was ambiguous. This result is similar to that of the meta-analysis conducted by Strozyk et al. in 2017, which included four RCTs [70]. Their study focused on assessing the efficacy of selenium supplementation in T2DM, and the results supported that selenium supplementation significantly improved insulin resistance. However, our study also focused on other cardiometabolic diseases, such as cardiovascular disease, and collected RCTs that have been updated recently. Therefore, 10 RCTs were included in this meta-analysis. The pooled results derived from the included studies are in line with those of previous animal studies [71–73], in that supplemental selenium therapy had a significantly protective anti-diabetic effect. Selenium nanoparticles (selenium-NPs) are a new supplemental form of selenium that is readily absorbed and has low toxicity. According to Abdulmalek et al. [74], treating diabetic rats with selenium-NPs (0.1 and 0.4 mg/kg) and/or metformin (100 mg/kg) separately or together, led to a remarkable decrease in FBG and insulin levels, suggesting that selenium and its derivatives play a significant role in the maintenance of glucose homeostasis. There is a lack of positive evidence regarding the effect of selenium supplementation on HbA1c. A cross-sectional study reported that obese participants with lower selenium intakes exhibited higher concentrations of HbA1c [75]. In this regard, the effect of selenium supplementation on long-term glycemic control deserves further attention.

The pooled results of this study also demonstrated that selenium supplementation increased HDL-C levels, but had little effect on other blood lipids. In addition, in the subgroup of disease types, we found that selenium supplementation increased HDL-C levels more significantly in participants with cardiovascular disease, followed by those with diabetes mellitus. HDL-C is essential for reverse cholesterol transport and has anti-inflammatory, anti-atherogenic, and anti-thrombotic effects [76]. The interaction of these properties contributes to the cardioprotective properties of HDL-C. Thus, appropriate selenium supplementation may contribute to the improvement of CMDs, particularly cardiovascular disease.

Some statistical heterogeneity was discovered in the pooled analysis of insulin levels, HOMA-IR, FPG, and HDL-C. The results were similar before and after sensitivity analysis, suggesting that the results were stable and reliable. In terms of insulin levels and HOMA-IR outcomes, we considered that Faghihi's study [48] administered inconsistent types or combinations of antidiabetic drugs at baseline, which might be an essential source of heterogeneity in the pooled analysis. In Faghihi's study [48], approximately 85.2% of participants in the control group received combined antidiabetic medication, compared to only 66.6% of participants in the selenium group. Additionally, subgroup analysis suggested that the inconsistency of participants' underlying diseases is a major source of heterogeneity in terms of FPG and HDL-C. In the future, more evaluable studies should be included in the analysis to better systematically evaluate the effectiveness of selenium supplementation in different types of CMDs.

This meta-analysis has several strengths. First, two reviewers independently used a comprehensive search strategy to identify all trials evaluating the effect of selenium supplementation in patients with CMDs and used standardized templates to extract data from included trials to guarantee data accuracy and reduce the impact of study design faults. Second, most included RCTs were high-quality with appropriate randomization, allocation concealment, and double-blinding methods. Third, no publication bias was found in this meta-analysis. Furthermore, we performed a thorough sensitivity analysis to examine the stability of our results. However, there are several limitations that should be considered. First, as the control group was placebo instead of different doses of selenium, the optimal dose of selenium supplementation cannot be accurately determined in this study. Second, the number of participants in the included RCTs was relatively small, with none having more than 100. Third, as only two studies evaluated HbA1c and there was high heterogeneity in the pooled analysis results, we could only perform a systematic review, which may affect the reliability and comprehensiveness of the evaluation of the efficacy of selenium supplementation on the HbA1c control. Fourth, although trials of the effects of selenium supplementation on CMDs have been conducted in countries other than Iran, trials in which other nutritional supplements were supplemented in addition to selenium in the intervention [77], and trials in which markers related to glucolipid metabolism were not reported [78] were excluded from this meta-analysis, and the final relevant included studies were all conducted in Iran, which may limit the generalizability. According to this, the recommendations of potential effects of selenium supplementation conclusions should be drawn with caution. Finally, we suggest that future clinical trials should pay more attention to the different doses of selenium supplements and the baseline level of serum selenium of CMDs patients, in order to further illuminate the therapeutic effects of selenium supplementation on CMDs.

5. Conclusions

This meta-analysis demonstrated that selenium supplementation may reduce the levels of serum insulin and HOMA-IR, and increase serum HDL-C levels, suggesting that selenium supplementation may be beneficial for reducing insulin resistance in patients with CMDs. This finding may provide support to prospective studies looking into selenium supplementation to manage cardiometabolic risk factors. However, in the case of selenium excess, the efficacy of selenium supplementation on glucolipid metabolism needs further evaluation.

Supplementary Materials: The following supporting information can be downloaded at: https://www.mdpi.com/article/10.3390/nu14224933/s1, Table S1. Search strategy for each electronic database; Figure S1. Sensitivity analysis of insulin levels; Figure S2. Sensitivity analysis of HOMA-IR; Figure S3. Forest plot of FPG; Figure S4. Sensitivity analysis of FPG; Figure S5. Forest plot of HDL-C; Figure S6. Sensitivity analysis of HDL-C; Figure S7. Forest plot of HDL-C after excluding Ghazi 2021 [42].

Author Contributions: Conceptualization and design, A.W., R.B. and Z.G.; software, J.O., Y.C. and Y.S.; validation, A.W. and R.B.; writing—original draft preparation, J.O., Y.C.; writing—review and editing, A.W. and R.B. All authors have read and agreed to the published version of the manuscript.

Funding: This research was funded by the National Natural Science Foundation of China (No. 82174216 AND No. 82004194), the Fundamental Research Funds for the Central public welfare research institutes (No. ZZ13-YQ-005, No. ZZ13-YQ-005-C1, No. ZZ13-YQ-014-C1 and No. ZZ13-YQ-014) and Scientific Fund of National Clinical Research Center for Chinese Medicine Cardiology (No. CMC2022009).

Institutional Review Board Statement: Not applicable.

Informed Consent Statement: Not applicable.

Data Availability Statement: Not applicable.

Conflicts of Interest: The authors declare no conflict of interest.

References

1. Guo, F.; Moellering, D.R.; Garvey, W.T. The progression of cardiometabolic disease: Validation of a new cardiometabolic disease staging system applicable to obesity. *Obesity* **2014**, *22*, 110–118. [CrossRef] [PubMed]
2. Hodkinson, A.; Kontopantelis, E.; Adeniji, C.; van Marwijk, H.; McMillian, B.; Bower, P.; Panagioti, M. Interventions Using Wearable Physical Activity Trackers Among Adults With Cardiometabolic Conditions: A Systematic Review and Meta-analysis. *JAMA Netw. Open* **2021**, *4*, e2116382. [CrossRef] [PubMed]
3. Jiao, Y.Y.; Wang, L.S.; Jiang, H.R.; Jia, X.F.; Wang, Z.H.; Wang, H.J.; Zhang, B.; Ding, G.G. Epidemiological characteristics and trends of cardiometabolic risk factors in residents aged 18–64 years in 15 provinces of China. *Zhonghua Liu Xing Bing Xue Za Zhi* **2022**, *43*, 1254–1261. [CrossRef] [PubMed]
4. Madlala, S.S.; Hill, J.; Kunneke, E.; Kengne, A.P.; Peer, N.; Faber, M. Dietary Diversity and its Association with Nutritional Status, Cardiometabolic Risk Factors and Food Choices of Adults at Risk for Type 2 Diabetes Mellitus in Cape Town, South Africa. *Nutrients* **2022**, *14*, 3191. [CrossRef] [PubMed]
5. Ghorbani, Z.; Kazemi, A.; Bartolomaeus, T.U.P.; Martami, F.; Noormohammadi, M.; Salari, A.; Löber, U.; Balou, H.A.; Forslund, S.K.; Mahdavi-Roshan, M. The effect of probiotic and synbiotic supplementation on lipid parameters among patients with cardiometabolic risk factors: A systematic review and meta-analysis of clinical trials. *Cardiovasc. Res.* 2022; *in press*. [CrossRef]
6. Jamilian, M.; Modarres, S.Z.; Siavashani, M.A.; Karimi, M.; Mafi, A.; Ostadmohammadi, V.; Asemi, Z. The Influences of Chromium Supplementation on Glycemic Control, Markers of Cardio-Metabolic Risk, and Oxidative Stress in Infertile Polycystic ovary Syndrome Women Candidate for In vitro Fertilization: A Randomized, Double-Blind, Placebo-Controlled Trial. *Biol. Trace Elem. Res.* **2018**, *185*, 48–55. [CrossRef]
7. Gil, Á.; Ortega, R.M. Introduction and Executive Summary of the Supplement, Role of Milk and Dairy Products in Health and Prevention of Noncommunicable Chronic Diseases: A Series of Systematic Reviews. *Adv. Nutr.* **2019**, *10*, S67–S73. [CrossRef]
8. Kim, Y.; Oh, Y.K.; Lee, J.; Kim, E. Could nutrient supplements provide additional glycemic control in diabetes management? A systematic review and meta-analysis of randomized controlled trials of as an add-on nutritional supplementation therapy. *Arch. Pharmacal Res.* **2022**, *45*, 185–204. [CrossRef]
9. Bitterli, C.; Bañuelos, G.S.; Schulin, R. Use of transfer factors to characterize uptake of selenium by plants. *J. Geochem. Explor.* **2010**, *107*, 206–216. [CrossRef]

10. Fordyce, F. Selenium Geochemistry and Health. *Ambio* **2007**, *36*, 94–97. [CrossRef]
11. Hadrup, N.; Ravn-Haren, G. Absorption, distribution, metabolism and excretion (ADME) of oral selenium from organic and inorganic sources: A review. *J. Trace Elem. Med. Biol.* **2021**, *67*, 126801. [CrossRef]
12. Labunskyy, V.M.; Hatfield, D.L.; Gladyshev, V.N. Selenoproteins: Molecular Pathways and Physiological Roles. *Physiol. Rev.* **2014**, *94*, 739–777. [CrossRef]
13. Fairweather-Tait, S.J.; Bao, Y.; Broadley, M.R.; Collings, R.; Ford, D.; Hesketh, J.E.; Hurst, R. Selenium in Human Health and Disease. *Antioxid. Redox Signal.* **2011**, *14*, 1337–1383. [CrossRef]
14. Handy, D.E.; Loscalzo, J. The role of glutathione peroxidase-1 in health and disease. *Free Radic. Biol. Med.* **2022**, *188*, 146–161. [CrossRef]
15. Hariharan, S.; Dharmaraj, S. Selenium and selenoproteins: It's role in regulation of inflammation. *Inflammopharmacology* **2020**, *28*, 667–695. [CrossRef]
16. Wang, J.; Yang, J.; Wang, C.; Zhao, Z.; Fan, Y. Systematic Review and Meta-Analysis of Oxidative Stress and Antioxidant Markers in Oral Lichen Planus. *Oxidative Med. Cell. Longev.* **2021**, *2021*, 9914652. [CrossRef]
17. Hasani, M.; Djalalinia, S.; Khazdooz, M.; Asayesh, H.; Zarei, M.; Gorabi, A.M.; Ansari, H.; Qorbani, M.; Heshmat, R. Effect of selenium supplementation on antioxidant markers: A systematic review and meta-analysis of randomized controlled trials. *Hormones* **2019**, *18*, 451–462. [CrossRef]
18. Hurrle, S.; Hsu, W.H. The etiology of oxidative stress in insulin resistance. *Biomed. J.* **2017**, *40*, 257–262. [CrossRef]
19. Singh, A.; Kukreti, R.; Saso, L.; Kukreti, S. Mechanistic Insight into Oxidative Stress-Triggered Signaling Pathways and Type 2 Diabetes. *Molecules* **2022**, *27*, 950. [CrossRef]
20. Kattoor, A.J.; Pothineni, N.V.K.; Palagiri, D.; Mehta, J.L. Oxidative Stress in Atherosclerosis. *Curr. Atheroscler. Rep.* **2017**, *19*, 42. [CrossRef]
21. Gunawardena, H.P.; Silva, R.; Sivakanesan, R.; Ranasinghe, P.; Katulanda, P. Poor Glycaemic Control Is Associated with Increased Lipid Peroxidation and Glutathione Peroxidase Activity in Type 2 Diabetes Patients. *Curr. Med. Res. Opin.* **2019**, *2019*, 9471697. [CrossRef] [PubMed]
22. Barchielli, G.; Capperucci, A.; Tanini, D. The Role of Selenium in Pathologies: An Updated Review. *Antioxidants* **2022**, *11*, 251. [CrossRef] [PubMed]
23. Wu, H.; Zhao, G.; Liu, S.; Zhang, Q.; Wang, P.; Cao, Y.; Wu, L. Supplementation with selenium attenuates autism-like behaviors and improves oxidative stress, inflammation and related gene expression in an autism disease model. *J. Nutr. Biochem.* **2022**, *107*, 109034. [CrossRef]
24. Brodin, O.; Eksborg, S.; Wallenberg, M.; Asker-Hagelberg, C.; Larsen, E.H.; Mohlkert, D.; Lenneby-Helleday, C.; Jacobsson, H.; Linder, S.; Misra, S.; et al. Pharmacokinetics and Toxicity of Sodium Selenite in the Treatment of Patients with Carcinoma in a Phase I Clinical Trial: The SECAR Study. *Nutrients* **2015**, *7*, 4978–4994. [CrossRef] [PubMed]
25. Nandi, A.; Yan, L.-J.; Jana, C.K.; Das, N. Role of Catalase in Oxidative Stress- and Age-Associated Degenerative Diseases. *Oxidative Med. Cell. Longev.* **2019**, *2019*, 9613090. [CrossRef]
26. Li, Z.-Y.; Xu, G.-B.; Xia, T.-A. Prevalence rate of metabolic syndrome and dyslipidemia in a large professional population in Beijing. *Atherosclerosis* **2006**, *184*, 188–192. [CrossRef]
27. Kohler, L.N.; Florea, A.; Kelley, C.P.; Chow, S.; Hsu, P.; Batai, K.; Saboda, K.; Lance, P.; Jacobs, E.T. Higher Plasma Selenium Concentrations Are Associated with Increased Odds of Prevalent Type 2 Diabetes. *J. Nutr.* **2018**, *148*, 1333–1340. [CrossRef]
28. Yin, T.; Zhu, X.; Xu, D.; Lin, H.; Lu, X.; Tang, Y.; Shi, M.; Yao, W.; Zhou, Y.; Zhang, H.; et al. The Association Between Dietary Antioxidant Micronutrients and Cardiovascular Disease in Adults in the United States: A Cross-Sectional Study. *Front. Nutr.* **2021**, *8*, 799095. [CrossRef]
29. Huang, J.; Xie, L.; Song, A.; Zhang, C. Selenium Status and Its Antioxidant Role in Metabolic Diseases. *Oxidative Med. Cell. Longev.* **2022**, *2022*, 7009863. [CrossRef]
30. Kamali, A.; Amirani, E.; Asemi, Z. Effects of Selenium Supplementation on Metabolic Status in Patients Undergoing for Coronary Artery Bypass Grafting (CABG) Surgery: A Randomized, Double-Blind, Placebo-Controlled Trial. *Biol. Trace Elem. Res.* **2019**, *191*, 331–337. [CrossRef]
31. Cardoso, B.R.; Braat, S.; Graham, R.M. Selenium Status Is Associated With Insulin Resistance Markers in Adults: Findings From the 2013 to 2018 National Health and Nutrition Examination Survey (NHANES). *Front. Nutr.* **2021**, *8*, 696024. [CrossRef]
32. Ju, W.; Ji, M.; Li, X.; Li, Z.; Wu, G.; Fu, X.; Yang, X.; Gao, X. Relationship between higher serum selenium level and adverse blood lipid profile. *Clin. Nutr.* **2018**, *37*, 1512–1517. [CrossRef]
33. Ju, W.; Li, X.; Li, Z.; Wu, G.R.; Fu, X.F.; Yang, X.M.; Zhang, X.Q.; Gao, X.B. The effect of selenium supplementation on coronary heart disease: A systematic review and meta-analysis of randomized controlled trials. *J. Trace Elem. Med. Biol.* **2017**, *44*, 8–16. [CrossRef]
34. Page, M.J.; McKenzie, J.E.; Bossuyt, P.M.; Boutron, I.; Hoffmann, T.C.; Mulrow, C.D.; Shamseer, L.; Tetzlaff, J.M.; Akl, E.A.; Brennan, S.E.; et al. The PRISMA 2020 statement: An updated guideline for reporting systematic reviews. *Syst. Rev.* **2021**, *10*, 89. [CrossRef]
35. Higgins, J.P.; Thomas, J.; Chandler, J.; Cumpston, M.; Li, T.; Page, M.J.; Welch, V.A. *Cochrane Handbook for Systematic Reviews of Interventions*; John Wiley & Sons: Hoboken, NJ, USA, 2019.

36. Yuan, D.; Sharma, H.; Krishnan, A.; Vangaveti, V.N.; Malabu, U.H. Effect of glucagon-like peptide 1 receptor agonists on albuminuria in adult patients with type 2 diabetes mellitus: A systematic review and meta-analysis. *Diabetes Obes. Metab.* **2022**, *24*, 1869–1881. [CrossRef]
37. Li, T.; Yu, T.; Hawkins, B.S.; Dickersin, K. Design, Analysis, and Reporting of Crossover Trials for Inclusion in a Meta-Analysis. *PLoS ONE* **2015**, *10*, e0133023. [CrossRef]
38. Cumpston, M.; Li, T.; Page, M.J.; Chandler, J.; Welch, V.A.; Higgins, J.P.; Thomas, J. Updated guidance for trusted systematic reviews: A new edition of the Cochrane Handbook for Systematic Reviews of Interventions. *Cochrane Database Syst. Rev.* **2019**, *10*, ED000142. [CrossRef]
39. Higgins, J.P.T.; Thompson, S.G.; Deeks, J.J.; Altman, D.G. Measuring inconsistency in meta-analyses. *BMJ* **2003**, *327*, 557–560. [CrossRef]
40. Akbari, M.; Moosazadeh, M.; Ghahramani, S.; Tabrizi, R.; Kolahdooz, F.; Asemi, Z.; Lankarani, K.B. High prevalence of hypertension among Iranian children and adolescents: A systematic review and meta-analysis. *J. Hypertens.* **2017**, *35*, 1155–1163. [CrossRef]
41. Salimian, M.; Soleimani, A.; Bahmani, F.; Tabatabaei, S.M.H.; Asemi, Z.; Talari, H.R. The effects of selenium administration on carotid intima-media thickness and metabolic status in diabetic hemodialysis patients: A randomized, double-blind, placebo-controlled trial. *Clin. Nutr. ESPEN* **2022**, *47*, 58–62. [CrossRef]
42. Ghazi, M.K.K.; Ghaffari, S.; Naemi, M.; Salehi, R.; Sadeghi, M.T.; Barati, M.; Shabestari, A.N.; Kafil, B.; Alamdari, N.M.; Soleimanzadeh, H.; et al. Effects of sodium selenite and selenium-enriched yeast on cardiometabolic indices of patients with atherosclerosis: A double-blind randomized clinical trial study. *J. Cardiovasc. Thorac. Res.* **2021**, *13*, 314–319. [CrossRef] [PubMed]
43. Najib, F.S.; Poordast, T.; Nia, M.R.; Dabbaghmanesh, M.H. Effects of selenium supplementation on glucose homeostasis in women with gestational diabetes mellitus: A randomized, controlled trial. *Int. J. Reprod. Biomed.* **2020**, *18*, 57–64. [CrossRef] [PubMed]
44. Raygan, F.; Behnejad, M.; Ostadmohammadi, V.; Bahmani, F.; Mansournia, M.A.; Karamali, F.; Asemi, Z. Selenium supplementation lowers insulin resistance and markers of cardio-metabolic risk in patients with congestive heart failure: A randomised, double-blind, placebo-controlled trial. *Br. J. Nutr.* **2018**, *120*, 33–40. [CrossRef] [PubMed]
45. Bahmani, F.; Kia, M.; Soleimani, A.; Asemi, Z.; Esmaillzadeh, A. Effect of Selenium Supplementation on Glycemic Control and Lipid Profiles in Patients with Diabetic Nephropathy. *Biol. Trace Elem. Res.* **2016**, *172*, 282–289. [CrossRef] [PubMed]
46. Farrokhian, A.; Bahmani, F.; Taghizadeh, M.; Mirhashemi, S.M.; Aarabi, M.; Raygan, F.; Aghadavod, E.; Asemi, Z. Selenium Supplementation Affects Insulin Resistance and Serum hs-CRP in Patients with Type 2 Diabetes and Coronary Heart Disease. *Horm. Metab. Res.* **2016**, *48*, 263–268. [CrossRef] [PubMed]
47. Asemi, Z.; Jamilian, M.; Mesdaghinia, E.; Esmaillzadeh, A. Effects of selenium supplementation on glucose homeostasis, inflammation, and oxidative stress in gestational diabetes: Randomized, double-blind, placebo-controlled trial. *Nutrition* **2015**, *31*, 1235–1242. [CrossRef]
48. Faghihi, T.; Radfar, M.; Barmal, M.; Amini, P.; Qorbani, M.; Abdollahi, M.; Larijani, B. A Randomized, Placebo-Controlled Trial of Selenium Supplementation in Patients With Type 2 Diabetes: Effects on Glucose Homeostasis, Oxidative Stress, and Lipid Profile. *Am. J. Ther.* **2014**, *21*, 491–495. [CrossRef]
49. Alizadeh, M.; Safaeiyan, A.; Ostadrahimi, A.; Estakhri, R.; Daneghian, S.; Ghaffari, A.; Gargari, B.P. Effect of L-Arginine and Selenium Added to a Hypocaloric Diet Enriched with Legumes on Cardiovascular Disease Risk Factors in Women with Central Obesity: A Randomized, Double-Blind, Placebo-Controlled Trial. *Ann. Nutr. Metab.* **2012**, *60*, 157–168. [CrossRef]
50. Yang, L.; Qi, M.; Du, X.; Xia, Z.; Fu, G.; Chen, X.; Liu, Q.; Sun, N.; Shi, C.; Zhang, R. Selenium concentration is associated with occurrence and diagnosis of three cardiovascular diseases: A systematic review and meta-analysis. *J. Trace Elem. Med. Biol.* **2022**, *70*, 126908. [CrossRef]
51. Xu, W.; Tang, Y.; Ji, Y.; Yu, H.; Li, Y.; Piao, C.; Xie, L. The association between serum selenium level and gestational diabetes mellitus: A systematic review and meta-analysis. *Diabetes Metab. Res. Rev.* **2022**, *38*, e3522. [CrossRef]
52. Fontenelle, L.C.; de Araújo, D.S.C.; da Cunha Soares, T.; Cruz, K.J.C.; Henriques, G.S.; do Nascimento Marreiro, D. Nutritional status of selenium in overweight and obesity: A systematic review and meta-analysis. *Clin. Nutr.* **2022**, *41*, 862–884. [CrossRef]
53. Kuria, A.; Tian, H.; Li, M.; Wang, Y.; Aaseth, J.O.; Zang, J.; Cao, Y. Selenium status in the body and cardiovascular disease: A systematic review and meta-analysis. *Crit. Rev. Food Sci. Nutr.* **2021**, *61*, 3616–3625. [CrossRef]
54. Ding, J.; Liu, Q.; Liu, Z.; Guo, H.; Liang, J.; Zhang, Y. Associations of the Dietary Iron, Copper, and Selenium Level With Metabolic Syndrome: A Meta-Analysis of Observational Studies. *Front. Nutr.* **2021**, *8*, 810494. [CrossRef]
55. Qiu, Z.; Geng, T.; Wan, Z.; Lu, Q.; Guo, J.; Liu, L.; Pan, A.; Liu, G. Serum selenium concentrations and risk of all-cause and heart disease mortality among individuals with type 2 diabetes. *Am. J. Clin. Nutr.* **2022**, *115*, 53–60. [CrossRef]
56. Díaz-Ruiz, M.; Martínez-Triguero, M.L.; López-Ruiz, A.; Fernández-de la Cruz, F.; Bañuls, C.; Hernández-Mijares, A. Metabolic disorders and inflammation are associated with familial combined hyperlipemia. *Clin. Chim. Acta* **2019**, *490*, 194–199. [CrossRef]
57. Püschel, G.P.; Klauder, J.; Henkel, J. Macrophages, Low-Grade Inflammation, Insulin Resistance and Hyperinsulinemia: A Mutual Ambiguous Relationship in the Development of Metabolic Diseases. *J. Clin. Med.* **2022**, *11*, 4358. [CrossRef]
58. DeFronzo, R.A.; Tripathy, D. Skeletal Muscle Insulin Resistance Is the Primary Defect in Type 2 Diabetes. *Diabetes Care* **2009**, *32* (Suppl. S2), S157–S163. [CrossRef]
59. Norouzi, S.; Adulcikas, J.; Sohal, S.S.; Myers, S. Zinc transporters and insulin resistance: Therapeutic implications for type 2 diabetes and metabolic disease. *J. Biomed. Sci.* **2017**, *24*, 87. [CrossRef]

60. Ogawa-Wong, A.N.; Berry, M.J.; Seale, L.A. Selenium and Metabolic Disorders: An Emphasis on Type 2 Diabetes Risk. *Nutrients* **2016**, *8*, 80. [CrossRef]
61. Ezaki, O. The insulin-like effects of selenate in rat adipocytes. *J. Biol. Chem.* **1990**, *265*, 1124–1128. [CrossRef]
62. Pasquier, A.; Vivot, K.; Erbs, E.; Spiegelhalter, C.; Zhang, Z.; Aubert, V.; Liu, Z.; Senkara, M.; Maillard, E.; Pinget, M.; et al. Lysosomal degradation of newly formed insulin granules contributes to β cell failure in diabetes. *Nat. Commun.* **2019**, *10*, 3312. [CrossRef] [PubMed]
63. Kitabayashi, N.; Nakao, S.; Mita, Y.; Arisawa, K.; Hoshi, T.; Toyama, T.; Ishii, K.-A.; Takamura, T.; Noguchi, N.; Saito, Y. Role of selenoprotein P expression in the function of pancreatic β cells: Prevention of ferroptosis-like cell death and stress-induced nascent granule degradation. *Free Radic. Biol. Med.* **2022**, *183*, 89–103. [CrossRef] [PubMed]
64. Tabrizi, R.; Akbari, M.; Moosazadeh, M.; Lankarani, K.B.; Heydari, S.T.; Kolahdooz, F.; Mohammadi, A.A.; Shabani, A.; Badehnoosh, B.; Jamilian, M.; et al. The Effects of Selenium Supplementation on Glucose Metabolism and Lipid Profiles Among Patients with Metabolic Diseases: A Systematic Review and Meta-Analysis of Randomized Controlled Trials. *Horm. Metab. Res.* **2017**, *49*, 826–830. [CrossRef] [PubMed]
65. Rayman, M.P.; Stranges, S. Epidemiology of selenium and type 2 diabetes: Can we make sense of it? *Free Radic. Biol. Med.* **2013**, *65*, 1557–1564. [CrossRef] [PubMed]
66. Wasserman, D.H. Four grams of glucose. *Am. J. Physiol. Endocrinol. Metab.* **2009**, *296*, E11–E21. [CrossRef] [PubMed]
67. Cherkas, A.; Holota, S.; Mdzinarashvili, T.; Gabbianelli, R.; Zarkovic, N. Glucose as a Major Antioxidant: When, What for and Why It Fails? *Antioxidants* **2020**, *9*, 140. [CrossRef] [PubMed]
68. Greenfield, M.S.; Doberne, L.; Kraemer, F.; Tobey, T.; Reaven, G. Assessment of Insulin Resistance with the Insulin Suppression Test and the Euglycemic Clamp. *Diabetes* **1981**, *30*, 387–392. [CrossRef]
69. Sherwani, S.I.; Khan, H.A.; Ekhzaimy, A.; Masood, A.; Sakharkar, M.K. Significance of HbA1c Test in Diagnosis and Prognosis of Diabetic Patients. *Biomark. Insights* **2016**, *11*, 95–104. [CrossRef]
70. Stróżyk, A.; Osica, Z.; Przybylak, J.D.; Kołodziej, M.; Zalewski, B.M.; Mrozikiewicz-Rakowska, B.; Szajewska, H. Effectiveness and safety of selenium supplementation for type 2 diabetes mellitus in adults: A systematic review of randomised controlled trials. *J. Hum. Nutr. Diet.* **2019**, *32*, 635–645. [CrossRef]
71. Liu, Y.; Zeng, S.; Liu, Y.; Wu, W.; Shen, Y.; Zhang, L.; Li, C.; Chen, H.; Liu, A.; Shen, L.; et al. Synthesis and antidiabetic activity of selenium nanoparticles in the presence of polysaccharides from *Catathelasma ventricosum*. *Int. J. Biol. Macromol.* **2018**, *114*, 632–639. [CrossRef]
72. Zeng, S.; Ke, Y.; Liu, Y.; Shen, Y.; Zhang, L.; Li, C.; Liu, A.; Shen, L.; Hu, X.; Wu, H.; et al. Synthesis and antidiabetic properties of chitosan-stabilized selenium nanoparticles. *Colloids Surf. B Biointerfaces* **2018**, *170*, 115–121. [CrossRef]
73. Oztürk, Z.; Gurpinar, T.; Vural, K.; Boyacıoglu, S.; Korkmaz, M.; Var, A. Effects of selenium on endothelial dysfunction and metabolic profile in low dose streptozotocin induced diabetic rats fed a high fat diet. *Biotech. Histochem.* **2015**, *90*, 506–515. [CrossRef]
74. Abdulmalek, S.A.; Balbaa, M. Synergistic effect of nano-selenium and metformin on type 2 diabetic rat model: Diabetic complications alleviation through insulin sensitivity, oxidative mediators and inflammatory markers. *PLoS ONE* **2019**, *14*, e0220779. [CrossRef]
75. Kazemi, A.; Shim, S.R.; Jamali, N.; Hassanzadeh-Rostami, Z.; Soltani, S.; Sasani, N.; Mohsenpour, M.A.; Firoozi, D.; Basirat, R.; Hosseini, R., et al. Comparison of nutritional supplements for glycemic control in type 2 diabetes: A systematic review and network meta-analysis of randomized trials. *Diabetes Res. Clin. Pract.* **2022**, *191*, 110037. [CrossRef]
76. Bhale, A.S.; Venkataraman, K. Leveraging knowledge of HDLs major protein ApoA1: Structure, function, mutations, and potential therapeutics. *Biomed. Pharmacother.* **2022**, *154*, 113634. [CrossRef]
77. Sena-Evangelista, K.C.M.; Pedrosa, L.F.C.; Paiva, M.S.M.O.; Dias, P.C.S.; Ferreira, D.Q.C.; Cozzolino, S.M.F.; Faulin, T.E.S.; Abdalla, D.S.P. The Hypolipidemic and Pleiotropic Effects of Rosuvastatin Are Not Enhanced by Its Association with Zinc and Selenium Supplementation in Coronary Artery Disease Patients: A Double Blind Randomized Controlled Study. *PLoS ONE* **2015**, *10*, e0119830. [CrossRef]
78. Gharipour, M.; Sadeghi, M.; Haghjooy-Javanmard, S.; Hamledari, H.; Khosravi, E.; Dianatkhah, M.; Vaseghi, G. Effects of selenium intake on the expression of prostaglandin-endoperoxide synthase 2 (cyclooxygenase-2) and matrix metallopeptidase-9 genes in the coronary artery disease: Selenegene study, a double-blind randomized controlled trial. *ARYA Atheroscler. J.* **2021**, *17*, 2093. [CrossRef]

Article

Dietary Inflammation Index and Its Association with Long-Term All-Cause and Cardiovascular Mortality in the General US Population by Baseline Glycemic Status

Sheng Yuan [1,2,†], Chenxi Song [1,2,†], Rui Zhang [1,2,†], Jining He [1,2] and Kefei Dou [1,2,*]

1. State Key Laboratory of Cardiovascular Disease, Beijing 102308, China; yuansheng@student.pumc.edu.cn (S.Y.); songchenxi@fuwaihospital.org (C.S.); zhangrui@fuwai.com (R.Z.); hejining95@163.com (J.H.)
2. Cardiometabolic Medicine Center, Department of Cardiology, Fuwai Hospital, National Center for Cardiovascular Diseases, Chinese Academy of Medical Sciences and Peking Union Medical College, Beijing 100037, China

* Correspondence: drdoukefei@126.com

† These authors contributed equally to this work.

Abstract: Dietary inflammatory potential has been proven to be correlated with the incidence of diabetes and cardiovascular diseases. However, the evidence regarding the impact of dietary inflammatory patterns on long-term mortality is scarce. This cohort study aims to investigate the dietary inflammatory pattern of the general US individuals by baseline glycemic status and to estimate its association with long-term mortality. A total of 20,762 general American adults with different glycemic statuses from the National Health and Nutrition Examination Survey were included. We extracted 24-h dietary information, and the dietary inflammatory index (DII) was calculated. The outcomes were defined as 5-year all-cause and cardiovascular mortality. Compared with the normoglycemia group, individuals with prediabetes and type 2 diabetes had higher DII scores (overall weighted $p < 0.001$). Compared with low DII scores, participants with high DII scores were at a higher risk of long-term all-cause mortality (HR: 1.597, 95% CI: 1.370, 1.861; $p < 0.001$) and cardiovascular mortality (HR: 2.036, 95% CI: 1.458, 2.844; $p < 0.001$). The results were stable after adjusting for potential confounders. Moreover, the prognostic value of DII for long-term all-cause mortality existed only in diabetic individuals but not in the normoglycemia or prediabetes group (p for interaction = 0.006). In conclusion, compared to the normoglycemia or prediabetes groups, participants with diabetes had a higher DII score, which indicates a greater pro-inflammatory potential. Diabetic individuals with higher DII scores were at a higher risk of long-term all-cause and cardiovascular mortality.

Keywords: diabetes mellitus; prediabetes; dietary inflammation index; nutrition; inflammatory diet; NHANES

Citation: Yuan, S.; Song, C.; Zhang, R.; He, J.; Dou, K. Dietary Inflammation Index and Its Association with Long-Term All-Cause and Cardiovascular Mortality in the General US Population by Baseline Glycemic Status. *Nutrients* **2022**, *14*, 2556. https://doi.org/10.3390/nu14132556

Academic Editor: Giuseppe Della Pepa

Received: 2 June 2022
Accepted: 16 June 2022
Published: 21 June 2022

Publisher's Note: MDPI stays neutral with regard to jurisdictional claims in published maps and institutional affiliations.

Copyright: © 2022 by the authors. Licensee MDPI, Basel, Switzerland. This article is an open access article distributed under the terms and conditions of the Creative Commons Attribution (CC BY) license (https://creativecommons.org/licenses/by/4.0/).

1. Introduction

Globally, the number of patients with diabetes and its devastating complications is increasing persistently in the past three decades, which is a major health threat to both developed and developing countries [1,2]. Due to the damage to the vascular smooth muscle cell and endothelial cell function [3], vascular diabetic complications cover almost all types of blood vessels and contribute to most of the mortality, hospitalization, and morbidity in patients with diabetes [4,5]. Obesity, decreased physical activity, population aging, and energy-dense diets are the primary causes of the rising diabetes rate [6]. Among those risk factors, the relationship between diabetes and nutrition or diet has received considerable attention [7–10].

Chronic inflammation plays a significant role in the etiology of diabetes [11,12]. Diet may interfere with the development of diabetes, which may be achieved through the influence of chronic inflammation. Many studies have demonstrated the correlation between pro-inflammatory food and diabetic risk [13,14]. A cross-sectional study of diabetes-free women revealed that red meat consumption was associated with elevated plasma inflammatory factors, fasting insulin, and glycated hemoglobin [13]. Moreover, an increasing number of studies have found that the Mediterranean diet, which was proven to have an anti-inflammatory effect [15–17], was associated with a lower diabetic risk [18–21].

The dietary inflammatory index (DII) was a dietary assessment tool developed based on the summary of published literature and aimed to estimate the inflammatory potential of an individual's diet [22]. A high DII score, which was associated with elevated inflammatory markers, such as C-reactive protein (CRP), indicates a pro-inflammation diet and has been reported to be correlated with an increased risk of obesity, type 2 diabetes, and cardiovascular diseases [23–27]. Moreover, populations with higher DII scores were proven to have higher cardiovascular mortality [28,29]. However, currently, evidence about the relationship between DII and long-term mortality of subjects with different glycemic statuses is scarce. Therefore, our study aims to investigate the long-term prognostic value of DII among participants with normoglycemia, prediabetes, and type 2 diabetes, which may contribute to precise prognosis prediction and diabetes management.

2. Materials and Methods

2.1. Study Population

This cohort study was conducted following the Strengthening the Reporting of Observational Studies in Epidemiology (STROBE) reporting guideline for cohort studies [30]. The participants included in our analysis were extracted from the National Health and Nutrition Examination Survey (NHANES), a periodically conducted survey that obtains nationally-representative samples of the general Americans with a complex, multistage probability design [31]. In this study, we extracted participants from the 2007–2014 cycle. Adults with complete 24-h dietary data were included. The exclusion criteria included: (1) age <20 years old; (2) participants with pregnancy; and (3) participants without endpoint information.

2.2. Dietary Information

Dietary information was extracted from NHANES, which was collected through 24-h dietary recall interviews in the mobile examination center and was validated by the Nutrition Methodology Working Group [32]. Following the DII calculating protocol reported by N. Shivappa et al. [22], 28 food parameters in NHANES were used to calculate the DII, including carbohydrates, protein, total fat, alcohol, fiber, cholesterol, saturated fat, monounsaturated fatty acids (MUFAs), polyunsaturated fatty acids (PUFAs), niacin, vitamin A, thiamin, riboflavin, vitamin B6, vitamin B12, vitamin C, vitamin D, vitamin E, Fe, Mg, Zinc, Selenium, folic acid, β-carotene, caffeine, energy, n-3 fatty acids, and n-6 fatty acids. Previous studies have reported that DII calculated based on less than 30 food parameters kept its predictive ability [33,34].

A lower negative DII score suggests an anti-inflammatory effect, while a higher positive DII score means a pro-inflammatory effect of diet. According to the methods of N. Shivappa et al. [22], the DII calculation should be standardized to a world database that contains standard mean and standard deviation for each food parameter. The database was constructed by examining the relationship between parameters, including food components, and inflammation, in 1943 published articles. A parameter with proof of anti-inflammation effect would obtain a score of "−1", while a food parameter would receive a score of "+1" if it was reported to be correlated with a reduced level of anti-inflammatory cytokines or increased level of proinflammatory cytokines.

These values were further weighted according to the study design. For each included parameter, we first extracted the individualized consumption value and then subtracted it from the standard mean and divided this value by the standard deviation. To minimize the effect of right skewing, these values were converted to a centered percentile score. To achieve a symmetrical distribution with values centered around 0 and bounded between

−1 and +1, each percentile score was doubled, and then we subtracted "1". This centered percentile value for each food parameter was then multiplied by its corresponding inflammatory effect score to obtain the DII score for each food component. Finally, 28 food-specific DII scores were summed to create the overall DII score for each participant.

2.3. Diseases and Endpoint Definitions

Type 2 diabetes was defined as a self-reported physician diagnosis of diabetes, glycated hemoglobin A_{1c} (HbA_{1c}) $\geq$ 6.5%, fasting glucose $\geq$ 7.8 mmol/L, or use of insulin or oral hypoglycemic medication. Prediabetes was defined as HbA1c 5.7%–6.4% (39–46 mmol/mol) or impaired fasting glucose (IFG) [fasting plasma glucose (FPG): 110–125 mg/dL (6.1–6.9 mmol/L)], or impaired glucose tolerance (IGT) [Oral glucose tolerance test 2-h glucose value $\geq$ 140 mg/dL (7.8 mmol/L) but < 200 mg/dL (11.1 mmol/L) and FPG < 126 (7.0 mmol/L)]. Hypertension was diagnosed as a self-reported physician diagnosis of hypertension, use of antihypertensive drugs, or systolic blood pressure $\geq$ 140 mmHg and/or diastolic blood pressure $\geq$ 90 mmHg (at least three times).

Participants who met at least one of the following criteria were diagnosed with hyperlipidemia: (1) elevated total cholesterol $\geq$ 200 mg/dL (5.18 mmol/L); (2) high triglyceride level ($\geq$150 mg/dL); (3) low density lipoprotein-cholesterol (LDL-c) $\geq$ 130 mg/dL (3.37 mmol/L); (4) high density lipoprotein-cholesterol (HDL-c) < 40 mg/dL (1.04 mmol/L) in men and 50 mg/dL (1.30 mmol/L) in women; and (5) use of cholesterol-lowering drugs. We set the time of follow-up time as 5 years. The primary endpoint of follow-up was all-cause death, which was extracted from the records of the National Death Index (NDI). The secondary endpoint was cardiovascular death, which was defined according to the International Classification of Diseases-10 codes (I00–I09, I11, I13, I20–I51) and was also extracted from NDI.

2.4. Statistics

To represent the overall US population, all analyses incorporated oversampling, clustering, and stratification as recommended by the NHANES data analysis guideline [31]. Continuous variables are listed as the weighted mean and 95% confidence interval (CI), while categorical variables are presented as weighted proportions. Basic characteristics are compared by baseline glycemic status using the adjusted Wald test for continuous variables and Rao-Scott χ^2 test for categorical variables.

The weighted Cox proportional hazard regression models were adopted to assess the impact of DII on participants' long-term mortality, which were adjusted for age, sex, educational level, BMI, smoke, hypertension, hyperlipidemia, glycemic status, recreational activity, and alcohol consumption. In addition to estimating DII as a continuous variable, we equally classified participants into three groups: low DII, medium DII, and high DII. Similar Cox regression models as well as weighted Kaplan-Meier curves were adopted to estimate the correlation between all-cause and cardiovascular mortality and different DII groups.

Furthermore, to test whether the impact of DII on prognosis is different across patients with different glycemic statuses, the weighted Cox regression model and interaction p value were used to estimate the relationship between DII (continuous/categorical variable) and participants' long-term mortality in participants with normal glucose status, prediabetes, and type 2 diabetes. The regression model was adjusted for age, sex, educational level, BMI, smoke, hypertension, hyperlipidemia, recreational activity, and alcohol consumption.

All analyses were conducted by the R software (version 4.1.2, R Foundation for Statistical Computing, Vienna, Austria) and Stata (version 16.0, StataCorp, College Station, TX, USA). A two-sided p value < 0.05 was considered statistically significant.

3. Results

3.1. Basic Characteristics by Baseline Glycemic Status

Following the pre-specified inclusion and exclusion criteria, a total of 20,762 participants were included in our study, among which 3859 were diagnosed with type 2 diabetes, 5489 with prediabetes, and 11,417 with normal glucose status (Figure 1). Table 1 lists the comparison of basic characteristics by glycemic status. Many variables showed an increasing relationship among patients with normoglycemia, prediabetes, and type 2 diabetes, such as age, BMI, waist, systolic blood pressure, HbA_{1c}, and triglycerides, which may indicate a worse health status in patients with prediabetes or type 2 diabetes. Similarly, we also found that patients with abnormal glucose metabolism were more likely to have a combination of hypertension or hyperlipidemia.

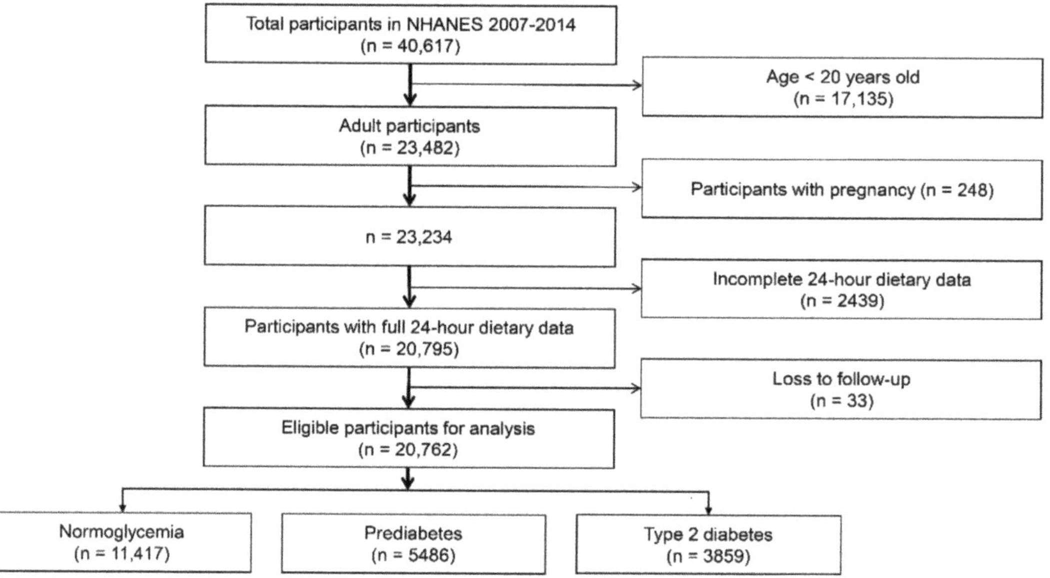

Figure 1. Flowchart of participant selection from NHANES database. NHANES: National Health and Nutrition Examination Survey.

Interestingly, compared with the normoglycemia group [113.94 (112.40,115.48) mg/dL], patients with prediabetes had a high level of LDL-c [122.04 (119.84, 124.24) mg/dL], while patients with type 2 diabetes had a better control of LDL-c [106.52 (103.97, 109.06) mg/dL]. As for the living habits, the percentage of former smokers was higher while the percentage of current smokers was lower in the diabetic population. The proportion of moderate or heavy drinking was also lower in participants with prediabetes or type 2 diabetes. Moreover, participants with type 2 diabetes were less likely to participate in recreational activity.

3.2. Comparison of DII Score by Baseline Glycemic Status

Compared with the normoglycemia group (0.883, 95% CI: 0.793, 0.973), participants with prediabetes (1.081, 95% CI: 0.981, 1.181) and type 2 diabetes (1.249, 95% CI: 1.151, 1.346) had higher DII scores (overall weighted $p < 0.001$). Figure 2 presents the distribution of DII scores among three groups. The proportion of high DII scores was higher in participants with prediabetes or type 2 diabetes. Moreover, we compared the component of DII scores among the three groups to find the main cause of the difference.

Table 1. Basic characteristics of participants by baseline glycemic status.

	Total (n = 20,762)	Normoglycemia (n = 11,417)	Prediabetes (n = 5486)	Type 2 Diabetes (n = 3859)	p
Age (years)	47.35 (46.78, 47.92)	42.06 (41.37,42.76)	54.20 (53.59,54.81)	59.58 (58.98,60.18)	<0.001
Male	48.70	48.62	47.86	50.36	0.250
BMI (kg/m^2)	28.81 (28.64, 28.93)	27.47 (27.28,27.65)	30.01 (29.71,30.31)	32.90 (32.52,33.27)	<0.001
Waist (cm)	98.62 (98.18, 99.05)	94.80 (94.26, 95.33)	102.14 (101.56,102.72)	109.93 (109.15,110.71)	<0.001
Ethnicity					<0.001
Non-Hispanic white	67.92	70.28	65.65	61.06	
Non-Hispanic black	11.34	9.53	13.75	15.38	
Mexican American	8.45	8.22	8.28	9.79	
Other Hispanic	5.47	5.45	5.17	6.09	
Other race	6.83	6.52	7.15	7.68	
Educational level					<0.001
Less than high school	17.45	14.40	20.29	26.48	
High school or equivalent	22.66	20.91	25.23	26.16	
College or above	59.59	64.69	54.48	47.36	
Smoke					<0.001
Never smoker	54.50	57.28	50.67	48.49	
Former smoker	24.42	21.20	27.00	34.67	
Current smoker	21.07	21.52	22.33	16.83	
Moderate or heavy drink	36.25	42.27	28.96	21.47	<0.001
Recreational activity					<0.001
No	46.62	40.55	52.41	64.24	
Moderate	28.56	27.45	31.72	28.10	
Vigorous	24.83	32.00	15.87	7.66	
SBP (mmHg)	122.42 (121.95,122.89)	119.21 (118.76,119.67)	126.19 (125.43,126.95)	130.57 (129.56,131.58)	<0.001
DBP (mmHg)	70.93 (70.49,71.37)	70.88 (70.39,71.38)	71.81 (71.32,72.30)	69.56 (68.79,70.33)	<0.001
Hypertension	36.67	25.62	47.14	68.96	<0.001
Hyperlipidemia	71.07	62.55	82.76	88.60	<0.001
HbA1c (%)	5.62 (5.60,5.64)	5.23 (5.22,5.24)	5.77 (5.76,5.79)	7.08 (7.01,7.15)	<0.001
Total cholesterol (mg/dL)	195.10 (194.12,196.08)	193.73 (192.50,194.96)	202.88 (201.21,204.55)	187.79 (185.72,189.85)	<0.001
HDL-c (mg/dL)	52.74 (52.32, 53.16)	54.50 (54.01,55.00)	51.32 (50.65,52.00)	47.11 (46.42,47.80)	<0.001
LDL-c (mg/dL)	115.12 (113.92,116.32)	113.94 (112.40,115.48)	122.04 (119.84,124.24)	106.52 (103.97,109.06)	<0.001
Triglycerides (mg/dL)	129.91 (126.76,133.06)	113.14 (109.94,116.35)	139.14 (133.90,144.38)	168.63 (159.26,178.00)	<0.001

BMI: body mass index; DBP: diastolic blood pressure; DII: dietary inflammatory index; HDL-c: high-density lipoprotein cholesterol; LDL-c: low-density lipoprotein cholesterol; and SBP: systolic blood pressure.

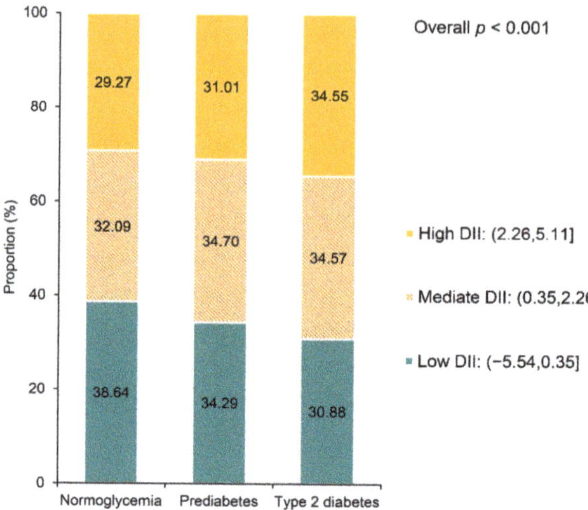

Figure 2. DII distribution by baseline glycemic status. DII: dietary inflammatory index.

Participants with type 2 diabetes had higher scores in alcohol, fiber, MUFA, PUFA, niacin, thiamin, riboflavin, vitamin B6, vitamin C, vitamin E, Mg, Zinc, Selenium, folic acid, N-3 fatty acids, and N-6 fatty acids (Figure 3, Table 2). We also noticed lower scores of participants with type 2 diabetes in certain components, such as carbohydrates, protein, total fat, saturated fat, vitamin B12, Fe, and energy. When compared to the normoglycemia group, the DII component scores remained consistent between participants with prediabetes and type 2 diabetes but to a lesser extent in the former.

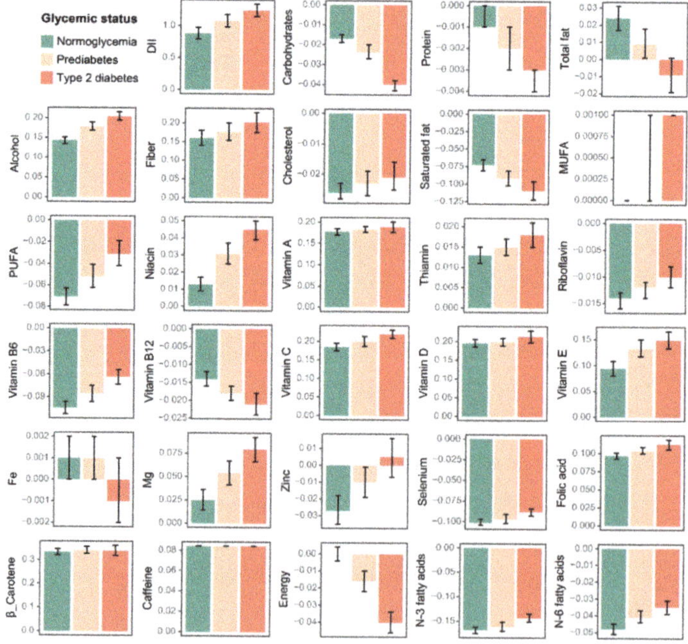

Figure 3. Comparison of the DII component scores by baseline glycemic status. Data are presented as the weighted mean value and 95%CI. DII: dietary inflammatory index; MUFA: monounsaturated fatty acids; and PUFA: polyunsaturated fatty acids.

Table 2. Comparison of the components of DII by baseline glycemic status.

	Total (n = 20,762)	Normoglycemia (n = 11,417)	Prediabetes (n = 5486)	Type 2 diabetes (n = 3859)	p
DII	0.980 (0.908, 1.052)	0.883 (0.793, 0.973)	1.081 (0.981, 1.181)	1.249 (1.151, 1.346)	<0.001
Carbohydrates	−0.022 (−0.023, −0.020)	−0.017 (−0.019, −0.015)	−0.024 (−0.027, −0.020)	−0.040 (−0.043, −0.038)	<0.001
Protein	−0.001 (−0.002, −0.001)	−0.001 (−0.001, 0.000)	−0.002 (−0.003, −0.001)	−0.003 (−0.004, −0.003)	<0.001
Total fat	0.016 (0.011, 0.021)	0.024 (0.017, 0.031)	0.009 (0.001, 0.018)	−0.009 (−0.019, 0.001)	<0.001
Alcohol	0.161 (0.154, 0.168)	0.144 (0.135, 0.152)	0.180 (0.169, 0.190)	0.205 (0.195, 0.216)	<0.001
Fiber	0.170 (0.154, 0.186)	0.160 (0.140, 0.180)	0.177 (0.154, 0.201)	0.202 (0.175, 0.228)	0.025
Cholesterol	−0.024 (−0.026, −0.022)	−0.026 (−0.028, −0.023)	−0.023 (−0.027, −0.019)	−0.021 (−0.025, −0.016)	0.2852
Saturated fat	−0.082 (−0.089, −0.075)	−0.073 (−0.081, −0.065)	−0.092 (−0.102, −0.081)	−0.109 (−0.122, −0.096)	<0.001
MUFA	0.000 (0.000, 0.001)	0.000 (0.000, 0.000)	0.000 (0.000, 0.001)	0.001 (0.001, 0.001)	0.001
PUFA	−0.061 (−0.067, −0.055)	−0.071 (−0.079, −0.063)	−0.052 (−0.062, −0.041)	−0.031 (−0.042, −0.019)	<0.001
Niacin	0.022 (0.019, 0.025)	0.013 (0.009, 0.017)	0.031 (0.025, 0.037)	0.045 (0.039, 0.050)	<0.001
Vitamin A	0.180 (0.174, 0.187)	0.178 (0.170, 0.185)	0.183 (0.176, 0.190)	0.189 (0.177, 0.201)	0.170
Thiamin	0.014 (0.012, 0.015)	0.013 (0.011, 0.015)	0.015 (0.013, 0.017)	0.018 (0.015, 0.021)	0.020
Riboflavin	−0.013 (−0.014, −0.012)	−0.014 (−0.016, −0.013)	−0.012 (−0.014, −0.011)	−0.010 (−0.012, −0.008)	0.001
Vitamin B6	−0.094 (−0.100, −0.088)	−0.104 (−0.112, −0.096)	−0.085 (−0.096, −0.075)	−0.063 (−0.073, −0.054)	<0.001
Vitamin B12	−0.016 (−0.017, −0.014)	−0.014 (−0.016, −0.012)	−0.018 (−0.020, −0.016)	−0.021 (−0.024, −0.018)	<0.001
Vitamin C	0.193 (0.185, 0.202)	0.185 (0.174, 0.195)	0.200 (0.187, 0.213)	0.220 (0.209, 0.231)	<0.001
Vitamin D	0.198 (0.191, 0.206)	0.195 (0.185, 0.205)	0.199 (0.188, 0.209)	0.213 (0.197, 0.229)	0.174
Vitamin E	0.111 (0.101, 0.121)	0.094 (0.080, 0.108)	0.133 (0.117, 0.150)	0.149 (0.133, 0.166)	<0.001
Fe	0.001 (0.000, 0.002)	0.001 (0.000, 0.002)	0.001 (0.000, 0.002)	−0.001 (−0.002, 0.001)	0.027
Mg	0.040 (0.031, 0.049)	0.025 (0.014, 0.036)	0.054 (0.041, 0.067)	0.079 (0.066, 0.092)	<0.001
Zinc	−0.018 (−0.025, −0.011)	−0.027 (−0.035, −0.018)	−0.010 (−0.019, −0.001)	0.005 (−0.007, 0.016)	<0.001
Selenium	−0.098 (−0.101, −0.096)	−0.101 (−0.104, −0.097)	−0.097 (−0.102, −0.091)	−0.088 (−0.093, −0.084)	<0.001
Folic acid	0.101 (0.098, 0.104)	0.097 (0.093, 0.101)	0.104 (0.100, 0.109)	0.113 (0.106, 0.119)	<0.001
β-Carotene	0.337 (0.326, 0.348)	0.335 (0.321, 0.348)	0.342 (0.327, 0.357)	0.340 (0.319, 0.361)	0.739
Caffeine	0.084 (0.084, 0.084)	0.034 (0.084, 0.084)	0.084 (0.084, 0.084)	0.084 (0.084, 0.084)	0.367
Energy	−0.009 (−0.012, −0.006)	0.000 (−0.004, 0.004)	−0.016 (−0.022, −0.010)	−0.040 (−0.046, −0.034)	<0.001
N-3 fatty acids	−0.163 (−0.168, −0.158)	−0.168 (−0.175, −0.162)	−0.161 (−0.170, −0.152)	−0.143 (−0.151, −0.135)	<0.001
N-6 fatty acids	−0.044 (−0.047, −0.042)	−0.048 (−0.051, −0.045)	−0.041 (−0.044, −0.037)	−0.035 (−0.039, −0.031)	<0.001

The data are presented as the mean and 95% confidence interval. DII: Dietary inflammation index; MUFA: monounsaturated fatty acids; PUFA: polyunsaturated fatty acids.

3.3. Association between Dietary Inflammation and Long-Term Mortality

The overall weighted 5-year all-cause mortality was 4.56%, and the weighted 5-year cardiovascular mortality was 1.17%. The Cox regression models revealed that higher DII scores were associated with higher long-term all-cause mortality (HR per 1 score increase: 1.105, 95% CI: 1.065, 1.147; $p < 0.001$) and cardiovascular mortality (HR per 1 score increase: 1.172, 95% CI: 1.092, 1.258; $p < 0.001$) (Table S1 in the Supplementary Materials).

The association was stable after adjusting for age, sex, educational level, BMI, smoke, hypertension, hyperlipidemia, glycemic status, recreational activity, and alcohol consumption. Compared with participants with low DII scores, participants with mediate or high DII scores had higher risk of all-cause death (Mediate DII: adjusted HR: 1.181, 95% CI: 1.009, 1.381; $p = 0.038$; high DII: adjusted HR: 1.240, 95% CI: 1.053, 1.459; $p = 0.010$) and cardiovascular death (adjusted HR: 1.442, 95% CI: 1.051, 1.979; $p = 0.023$; high DII: adjusted HR: 1.423, 95% CI: 1.006, 2.013; $p = 0.046$) (Figure 4, Table S1 in the Supplementary Materials).

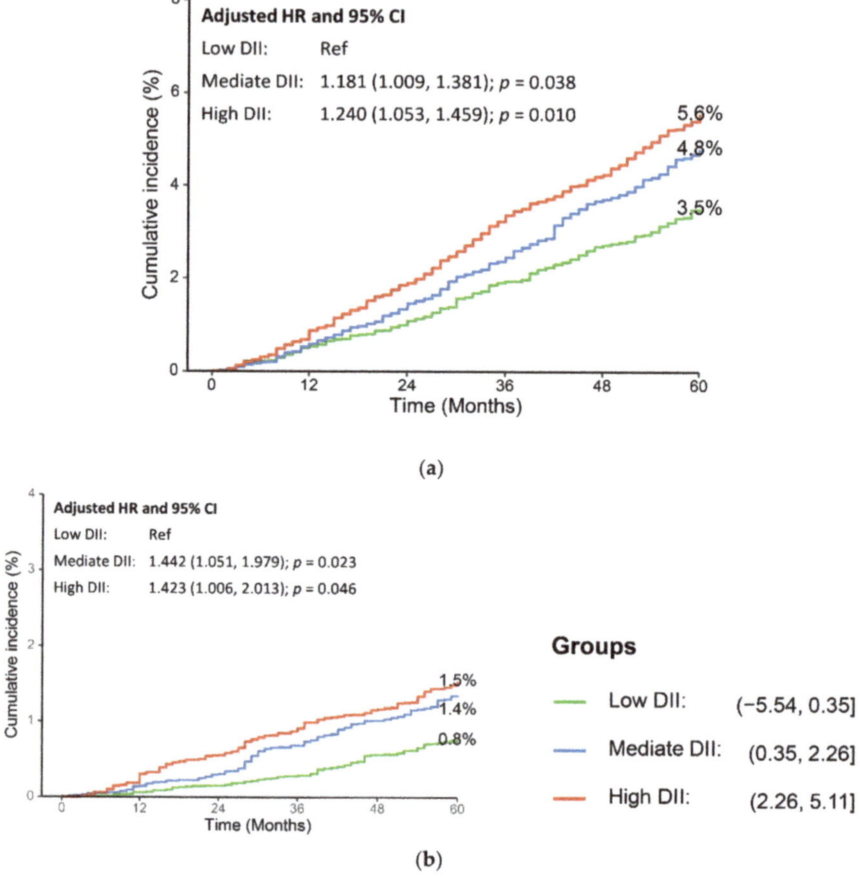

Figure 4. Association between DII scores and long-term (**a**) all-cause mortality and (**b**) cardiovascular mortality. HR was adjusted for age, sex, educational level, BMI, smoke, hypertension, hyperlipidemia, glycemic status, recreational activity, and alcohol consumption. CI: confidence interval; DII: dietary inflammatory index; HR: hazard ratio; and Ref: reference.

3.4. Dietary inflammation and Long-Term Mortality across Participants with Different Glycemic STATUSES

To estimate the impact of baseline glycemic status on the long-term prognostic value of DII, we performed adjusted Cox regression models in three groups: normoglycemia, prediabetes, and type 2 diabetes groups. As shown in Table 3, the association between DII scores and 5-year all-cause mortality was only significant in participants with type 2 diabetes (adjusted HR per 1 score increase 1.083, 95% CI: 1.014, 1.156; $p = 0.017$). DII was a better long-term all-cause mortality indicator in the type 2 diabetes group than was the normoglycemia or prediabetes group (p for interaction = 0.030).

Table 3. Association between DII and the long-term mortality of participants by baseline glycemic status.

	Normoglycemia (n = 11,417)	Prediabetes (n = 5486)	Type 2 Diabetes (n = 3859)	p for Interaction
All-cause mortality				
Continuous, per 1 score	1.004 (0.937, 1.075); $p = 0.919$	1.047 (0.967, 1.133); $p = 0.255$	1.083 (1.014, 1.156); $p = 0.017$	0.030
1st tertile (−5.54, 0.35]	Ref	Ref	Ref	0.006
2nd tertile (0.35, 2.26]	1.119 (0.834, 1.502); $p = 0.455$	0.878 (0.640, 1.204); $p = 0.418$	1.683 (1.266, 2.237); $p < 0.001$	
3rd tertile (2.26, 5.11]	1.011 (0.728, 1.404); $p = 0.947$	1.210 (0.848, 1.726); $p = 0.295$	1.626 (1.208, 2.188); $p = 0.001$	
Cardiovascular mortality				
Continuous, per 1 score	1.102 (0.958, 1.269); $p = 0.175$	1.032 (0.900, 1.184); $p = 0.654$	1.104 (0.962, 1.265); $p = 0.158$	0.867
1st tertile (−5.54, 0.35]	Ref	Ref	Ref	0.455
2nd tertile (0.35, 2.26]	1.548 (0.978, 2.449); $p = 0.062$	0.986 (0.527, 1.845); $p = 0.965$	1.864 (1.036, 3.353); $p = 0.038$	
3rd tertile (2.26, 5.11]	1.476 (0.725, 3.007); $p = 0.283$	0.934 (0.501, 1.739); $p = 0.829$	1.980 (1.043, 3.761); $p = 0.037$	

The data are presented as the adjusted HR and 95% CI. The Cox regression models are adjusted for age, sex, body mass index, smoke, hypertension, educational level, hyperlipidemia, recreational activity, and moderate or heavy drinker.

When treated as a categorical variable, a high DII score of participants with type 2 diabetes was associated with higher 5-year all-cause (adjusted HR 1.626, 95% CI: 1.208, 2.188; $p = 0.001$) and cardiovascular mortality (adjusted HR 1.980, 95% CI: 1.043, 3.761; $p = 0.037$) compared with low DII score. Participants with mediate DII scores in the type 2 diabetes group had a similar risk of long-term mortality. However, there was no significant correlation between DII and long-term mortality in the normoglycemia and prediabetes groups. The superiority of DII's prognostic value for long-term all-cause mortality in the type 2 diabetes group over the normoglycemia or prediabetes group was robust. (Continuous DII: p for interaction = 0.030; categorical DII: p for interaction = 0.006)

4. Discussion

Our study included a total of 20,762 participants, which represented 218,988,071 of the general US population, and we found that prediabetic or diabetic participants had a more pro-inflammatory diet compared with the normoglycemia group. Participants with mediate or higher DII scores were at higher risk of long-term all-cause and cardiovascular mortality. The prognostic effect of DII was only significant in diabetic participants and not in the prediabetic or normoglycemia group.

Many studies have shown that certain diet patterns, such as advanced glycation end products (AGEs), antioxidant diet, and the Mediterranean diet, can affect the low-level inflammation or body composition, and thus influence the incidence and development of some chronic diseases [15,35,36]. Previous research has suggested that dietary patterns may influence the incidence of diabetes. An analysis of 200,727 US participants from three prospective cohort studies conducted over 20 years revealed that eating more healthy plant foods and eating fewer animal foods was associated with a 20% reduction in diabetic risk [37].

A 20-year prospective cohort of 70,991 women discovered that a higher anti-inflammatory diet (as measured by DII) was linked to a reduced risk of type 2 diabetes [26]. Our study confirmed this relationship and found a sequentially increasing DII score across the normoglycemia, prediabetes, and type 2 diabetes groups. Moreover, component analysis in our results revealed that participants with prediabetes or type 2 diabetes had higher scores in alcohol, fiber, MUFA, PUFA, niacin, thiamin, riboflavin, vitamin B6, vitamin C, vitamin E, Mg, Zinc, Selenium, folic acid, N-3 fatty acids, and N-6 fatty acids compared with participants with normoglycemia.

Interestingly, diabetic participants had lower scores in some key nutritional indicators, such as carbohydrates, protein, total fat, saturated fat, and energy. This dietary pattern may come from the active adjustment after the diagnosis of prediabetes or diabetes. Another study based on the NHANES database discovered that participants with diagnosed prediabetes or diabetes were more likely to be concerned about nutrition fact labels when making daily food purchases [38].

However, rather than simple calorie calculations, we should be concerned about the complex and long-term influences of different foods on health [39]. Nutrition science found that overall dietary patterns and specific foods, instead of single isolated nutrients were more important for cardiometabolic health [40,41]. In participants with prediabetes or diabetes, a shortage of vitamins, critical micronutrients, and unsaturated fatty acids, as shown in our study, may lead to poor health and disease progression, which requires attention in diabetic care.

Dietary patterns are linked not only to the occurrence of chronic diseases but also to disease prognosis. A meta-analysis of 14 research articles found that individuals in the highest DII group had a higher risk of cardiovascular disease incidence as well as cardiovascular mortality [42]. Park et al. estimated the relationship between dietary inflammatory potential and prognosis in participants with different metabolic phenotypes [34]. They included 3733 adults from the NHANES III database (1988–1994) and revealed that the DII score was correlated with elevated all-cause and cardiovascular mortality in individuals with metabolically unhealthy obesity, which has not been observed in metabolically healthy obese individuals.

The target population of our study consists of 20,762 participants who participated in the NHANES project in the near twenty years (2007–2014). Similarly, our results demonstrated that a higher DII score was associated with higher long-term all-cause and cardiovascular mortality in participants with type 2 diabetes. The correlation was not identified in the prediabetes or normoglycemia group. Our findings imply that dietary inflammatory potential has a major influence on the long-term prognosis of diabetic patients, a topic that requires further attention in diabetes management.

To our knowledge, this is the first work that compares the long-term prognostic value of DII in the general American participants by baseline glycemic status. There are several limitations to our study. First, DII was calculated from self-reported dietary data, and recall bias was inevitable. Secondary, we extracted the 24 h dietary information to represent the daily pattern, which may change over time. Second, the DII used in our study was calculated from 28 food parameters due to the limitation of dietary data in the NHANES database. Previous studies have reported that DII calculated based on less than 30 food parameters kept its predictive ability [33,34].

Thirdly, we discovered that participants with prediabetes had higher LDL-c levels than the normoglycemia group, whereas patients with type 2 diabetes had better LDL-c control. This phenomenon could be explained by the fact that people with diabetes are more likely to visit the hospital and undergo laboratory tests, allowing their complications, such as hyperlipidemia, to be better managed.

However, this is our hypothesis, and because therapy data is limited, a specific reason should be investigated in future research. Finally, although we adjusted the potential risk factors including age, sex, body mass index, smoke, hypertension, educational level, hyperlipidemia, glycemic status, recreational activity, and moderate or heavy drinker in the multivariable Cox regression models, cardiovascular pathology and medication therapy were not involved due to the limitation of database, which may have an important impact on the cardiovascular mortality.

5. Conclusions

Our study identified a more pro-inflammatory diet in the diabetic participants compared with the general Americans. Participants with a higher DII score were at higher risk of 5-year all-cause and cardiovascular mortality. The prognostic value of DII existed only in individuals with type 2 diabetes but not in the normoglycemia or prediabetes group. The result calls for a comprehensive assessment of the dietary inflammatory potential in diabetic patients. Moreover, whether an anti-inflammatory dietary adjustment can improve the long-term prognosis of diabetes should be assessed in future trials.

Supplementary Materials: The following supporting information can be downloaded at: https://www.mdpi.com/article/10.3390/nu14132556/s1, Table S1: Survival analysis of the relationship between DII scores and long-term mortality.

Author Contributions: Conceptualization, K.D. and S.Y.; methodology, C.S. and S.Y.; software, S.Y. and J.H.; validation, R.Z. and J.H.; formal analysis, S.Y.; investigation, C.S.; resources, R.Z.; data curation, R.Z.; writing—original draft preparation, S.Y.; writing—review and editing, C.S. and K.D.; visualization, S.Y.; supervision, K.D.; project administration, K.D.; funding acquisition, K.D. All authors have read and agreed to the published version of the manuscript.

Funding: This research was funded by the CAMS Innovation Fund for Medical Sciences (CIFMS), grant number 2021-I2M-1-008 and the National Natural Science Foundation of China, grant number 81870277. The APC was funded by the CAMS Innovation Fund for Medical Sciences (CIFMS), grant number 2021-I2M-1-008.

Institutional Review Board Statement: Not applicable.

Informed Consent Statement: Not applicable.

Data Availability Statement: All data are available at NHANES website https://www.cdc.gov/nchs/nhanes/index.htm (accessed on 1 June 2022).

Acknowledgments: The NHANES protocol was approved by the NCHS Research Ethics Review Board.

Conflicts of Interest: The authors declare no conflict of interest.

References

1. Cole, J.B.; Florez, J.C. Genetics of Diabetes Mellitus and Diabetes Complications. *Nat. Rev. Nephrol.* **2020**, *16*, 377–390. [CrossRef]
2. Zheng, Y.; Ley, S.H.; Hu, F.B. Global Aetiology and Epidemiology of Type 2 Diabetes Mellitus and Its Complications. *Nat. Rev. Endocrinol.* **2018**, *14*, 88–98. [CrossRef]
3. Beckman, J.A.; Creager, M.A. Vascular Complications of Diabetes. *Circ. Res.* **2016**, *118*, 1771–1785. [CrossRef]
4. Tseng, C.-H. Mortality and Causes of Death in a National Sample of Diabetic Patients in Taiwan. *Diabetes Care* **2004**, *27*, 1605–1609. [CrossRef]
5. Morrish, N.J.; Wang, S.L.; Stevens, L.K.; Fuller, J.H.; Keen, H. Mortality and Causes of Death in the WHO Multinational Study of Vascular Disease in Diabetes. *Diabetologia* **2001**, *44* (Suppl. S2), S14–S21. [CrossRef]
6. Chatterjee, S.; Khunti, K.; Davies, M.J. Type 2 Diabetes. *Lancet* **2017**, *389*, 2239–2251. [CrossRef]
7. Ley, S.H.; Hamdy, O.; Mohan, V.; Hu, F.B. Prevention and Management of Type 2 Diabetes: Dietary Components and Nutritional Strategies. *Lancet* **2014**, *383*, 1999–2007. [CrossRef]
8. Wang, D.D.; Hu, F.B. Precision Nutrition for Prevention and Management of Type 2 Diabetes. *Lancet Diabetes Endocrinol.* **2018**, *6*, 416–426. [CrossRef]

9. Forouhi, N.G.; Misra, A.; Mohan, V.; Taylor, R.; Yancy, W. Dietary and Nutritional Approaches for Prevention and Management of Type 2 Diabetes. *BMJ* **2018**, *361*, k2234. [CrossRef]
10. Magkos, F.; Hjorth, M.F.; Astrup, A. Diet and Exercise in the Prevention and Treatment of Type 2 Diabetes Mellitus. *Nat. Rev. Endocrinol.* **2020**, *16*, 545–555. [CrossRef]
11. Rohm, T.V.; Meier, D.T.; Olefsky, J.M.; Donath, M.Y. Inflammation in Obesity, Diabetes, and Related Disorders. *Immunity* **2022**, *55*, 31–55. [CrossRef]
12. Furman, D.; Campisi, J.; Verdin, E.; Carrera-Bastos, P.; Targ, S.; Franceschi, C.; Ferrucci, L.; Gilroy, D.W.; Fasano, A.; Miller, G.W.; et al. Chronic Inflammation in the Etiology of Disease across the Life Span. *Nat. Med.* **2019**, *25*, 1822–1832. [CrossRef]
13. Ley, S.H.; Sun, Q.; Willett, W.C.; Eliassen, A.H.; Wu, K.; Pan, A.; Grodstein, F.; Hu, F.B. Associations between Red Meat Intake and Biomarkers of Inflammation and Glucose Metabolism in Women. *Am. J. Clin. Nutr.* **2014**, *99*, 352–360. [CrossRef]
14. Schulze, M.B.; Hoffmann, K.; Manson, J.E.; Willett, W.C.; Meigs, J.B.; Weikert, C.; Heidemann, C.; Colditz, G.A.; Hu, F.B. Dietary Pattern, Inflammation, and Incidence of Type 2 Diabetes in Women. *Am. J. Clin. Nutr.* **2005**, *82*, 675–684. [CrossRef]
15. Mena, M.-P.; Sacanella, E.; Vazquez-Agell, M.; Morales, M.; Fitó, M.; Escoda, R.; Serrano-Martínez, M.; Salas-Salvadó, J.; Benages, N.; Casas, R.; et al. Inhibition of Circulating Immune Cell Activation: A Molecular Antiinflammatory Effect of the Mediterranean Diet. *Am. J. Clin. Nutr.* **2009**, *89*, 248–256. [CrossRef]
16. Chrysohoou, C.; Panagiotakos, D.B.; Pitsavos, C.; Das, U.N.; Stefanadis, C. Adherence to the Mediterranean Diet Attenuates Inflammation and Coagulation Process in Healthy Adults: The ATTICA Study. *J. Am. Coll. Cardiol.* **2004**, *44*, 152–158. [CrossRef]
17. Estruch, R.; Martínez-González, M.A.; Corella, D.; Salas-Salvadó, J.; Ruiz-Gutiérrez, V.; Covas, M.I.; Fiol, M.; Gómez-Gracia, E.; López-Sabater, M.C.; Vinyoles, E.; et al. Effects of a Mediterranean-Style Diet on Cardiovascular Risk Factors: A Randomized Trial. *Ann. Intern. Med.* **2006**, *145*, 1–11. [CrossRef]
18. Martín-Peláez, S.; Fito, M.; Castaner, O. Mediterranean Diet Effects on Type 2 Diabetes Prevention, Disease Progression, and Related Mechanisms: A Review. *Nutrients* **2020**, *12*, 2236. [CrossRef]
19. Salas-Salvadó, J.; Bulló, M.; Estruch, R.; Ros, E.; Covas, M.-I.; Ibarrola-Jurado, N.; Corella, D.; Arós, F.; Gómez-Gracia, E.; Ruiz-Gutiérrez, V.; et al. Prevention of Diabetes with Mediterranean Diets: A Subgroup Analysis of a Randomized Trial. *Ann. Intern. Med.* **2014**, *160*, 1–10. [CrossRef]
20. InterAct Consortium; Romaguera, D.; Guevara, M.; Norat, T.; Langenberg, C.; Forouhi, N.G.; Sharp, S.; Slimani, N.; Schulze, M.B.; Buijsse, B.; et al. Mediterranean Diet and Type 2 Diabetes Risk in the European Prospective Investigation into Cancer and Nutrition (EPIC) Study: The InterAct Project. *Diabetes Care* **2011**, *34*, 1913–1918. [CrossRef]
21. Esposito, K.; Maiorino, M.I.; Ceriello, A.; Giugliano, D. Prevention and Control of Type 2 Diabetes by Mediterranean Diet: A Systematic Review. *Diabetes Res. Clin. Pract.* **2010**, *89*, 97–102. [CrossRef]
22. Shivappa, N.; Steck, S.E.; Hurley, T.G.; Hussey, J.R.; Hébert, J.R. Designing and Developing a Literature-Derived, Population-Based Dietary Inflammatory Index. *Public Health Nutr.* **2014**, *17*, 1689–1696. [CrossRef]
23. Hariharan, R.; Odjidja, E.N.; Scott, D.; Shivappa, N.; Hébert, J.R.; Hodge, A.; de Courten, B. The Dietary Inflammatory Index, Obesity, Type 2 Diabetes, and Cardiovascular Risk Factors and Diseases. *Obes. Rev.* **2022**, *23*, e13349. [CrossRef]
24. Tyrovolas, S.; Koyanagi, A.; Kotsakis, G.A.; Panagiotakos, D.; Shivappa, N.; Wirth, M.D.; Hébert, J.R.; Haro, J.M. Dietary Inflammatory Potential Is Linked to Cardiovascular Disease Risk Burden in the US Adult Population. *Int. J. Cardiol.* **2017**, *240*, 409–413. [CrossRef]
25. Sen, S.; Rifas-Shiman, S.L.; Shivappa, N.; Wirth, M.D.; Hebert, J.R.; Gold, D.R.; Gillman, M.W.; Oken, E. Associations of Prenatal and Early Life Dietary Inflammatory Potential with Childhood Adiposity and Cardiometabolic Risk in Project Viva. *Pediatr. Obes.* **2018**, *13*, 292–300. [CrossRef]
26. Laouali, N.; Mancini, F.R.; Hajji-Louati, M.; El Fatouhi, D.; Balkau, B.; Boutron-Ruault, M.-C.; Bonnet, F.; Fagherazzi, G. Dietary Inflammatory Index and Type 2 Diabetes Risk in a Prospective Cohort of 70,991 Women Followed for 20 Years: The Mediating Role of BMI. *Diabetologia* **2019**, *62*, 2222–2232. [CrossRef]
27. Wang, Y.; Armijos, R.X.; Xun, P.; Weigel, M.M. Dietary Inflammatory Index and Cardiometabolic Risk in Ecuadorian Women. *Nutrients* **2021**, *13*, 2640. [CrossRef]
28. Okada, E.; Shirakawa, T.; Shivappa, N.; Wakai, K.; Suzuki, K.; Date, C.; Iso, H.; Hébert, J.R.; Tamakoshi, A. Dietary Inflammatory Index Is Associated with Risk of All-Cause and Cardiovascular Disease Mortality but Not with Cancer Mortality in Middle-Aged and Older Japanese Adults. *J. Nutr.* **2019**, *149*, 1451–1459. [CrossRef]
29. Shivappa, N.; Steck, S.E.; Hussey, J.R.; Ma, Y.; Hebert, J.R. Inflammatory Potential of Diet and All-Cause, Cardiovascular, and Cancer Mortality in National Health and Nutrition Examination Survey III Study. *Eur. J. Nutr.* **2017**, *56*, 683–692. [CrossRef]
30. Von Elm, E.; Altman, D.G.; Egger, M.; Pocock, S.J.; Gøtzsche, P.C.; Vandenbroucke, J.P. STROBE Initiative The Strengthening the Reporting of Observational Studies in Epidemiology (STROBE) Statement: Guidelines for Reporting Observational Studies. *Ann. Intern. Med.* **2007**, *147*, 573–577. [CrossRef]
31. Centers for Disease Control and Prevention. National Center for Health Statistics NHANES Survey Methods and Analytic Guidelines. Available online: https://wwwn.cdc.gov/nchs/nhanes/analyticguidelines.aspx (accessed on 5 April 2022).
32. Plan and Operation of the Third National Health and Nutrition Examination Survey, 1988–1994. Series 1: Programs and Collection Procedures. *Vital Health Stat. 1* **1994**, *32*, 1–407.

33. Shivappa, N.; Steck, S.E.; Hurley, T.G.; Hussey, J.R.; Ma, Y.; Ockene, I.S.; Tabung, F.; Hébert, J.R. A Population-Based Dietary Inflammatory Index Predicts Levels of C-Reactive Protein in the Seasonal Variation of Blood Cholesterol Study (SEASONS). *Public Health Nutr.* **2014**, *17*, 1825–1833. [CrossRef] [PubMed]
34. Park, Y.-M.M.; Choi, M.K.; Lee, S.-S.; Shivappa, N.; Han, K.; Steck, S.E.; Hébert, J.R.; Merchant, A.T.; Sandler, D.P. Dietary Inflammatory Potential and Risk of Mortality in Metabolically Healthy and Unhealthy Phenotypes among Overweight and Obese Adults. *Clin. Nutr.* **2019**, *38*, 682–688. [CrossRef] [PubMed]
35. Garay-Sevilla, M.E.; Rojas, A.; Portero-Otin, M.; Uribarri, J. Dietary AGEs as Exogenous Boosters of Inflammation. *Nutrients* **2021**, *13*, 2802. [CrossRef] [PubMed]
36. Rashidmayvan, M.; Sharifan, P.; Darroudi, S.; Saffar Soflaei, S.; Salaribaghoonabad, R.; Safari, N.; Yousefi, M.; Honari, M.; Ghazizadeh, H.; Ferns, G.; et al. Association between Dietary Patterns and Body Composition in Normal-Weight Subjects with Metabolic Syndrome. *J. Diabetes Metab. Disord.* **2022**, *21*, 735–741. [CrossRef]
37. Satija, A.; Bhupathiraju, S.N.; Rimm, E.B.; Spiegelman, D.; Chiuve, S.E.; Borgi, L.; Willett, W.C.; Manson, J.E.; Sun, Q.; Hu, F.B. Plant-Based Dietary Patterns and Incidence of Type 2 Diabetes in US Men and Women: Results from Three Prospective Cohort Studies. *PLoS Med.* **2016**, *13*, e1002039. [CrossRef]
38. An, R. Diabetes Diagnosis and Nutrition Facts Label Use among US Adults, 2005–2010. *Public Health Nutr.* **2016**, *19*, 2149–2156. [CrossRef]
39. Mozaffarian, D. Dietary and Policy Priorities for Cardiovascular Disease, Diabetes, and Obesity: A Comprehensive Review. *Circulation* **2016**, *133*, 187–225. [CrossRef]
40. Jacobs, D.R.; Tapsell, L.C. Food, Not Nutrients, Is the Fundamental Unit in Nutrition. *Nutr. Rev.* **2007**, *65*, 439–450. [CrossRef]
41. Mozaffarian, D.; Ludwig, D.S. Dietary Guidelines in the 21st Century—A Time for Food. *JAMA* **2010**, *304*, 681–682. [CrossRef]
42. Shivappa, N.; Godos, J.; Hébert, J.R.; Wirth, M.D.; Piuri, G.; Speciani, A.F.; Grosso, G. Dietary Inflammatory Index and Cardiovascular Risk and Mortality-A Meta-Analysis. *Nutrients* **2018**, *10*, E200. [CrossRef] [PubMed]

 *nutrients* 

Article

Dietary Plant Sterols and Phytosterol-Enriched Margarines and Their Relationship with Cardiovascular Disease among Polish Men and Women: The WOBASZ II Cross-Sectional Study

Anna Maria Witkowska [1,*,†], Anna Waśkiewicz [2,†], Małgorzata Elżbieta Zujko [1], Alicja Cicha-Mikołajczyk [2], Iwona Mirończuk-Chodakowska [1] and Wojciech Drygas [2,3]

1 Department of Food Biotechnology, Faculty of Health Sciences, Medical University of Bialystok, Szpitalna 37, 15-295 Bialystok, Poland; malgorzata.zujko@umb.edu.pl (M.E.Z.); iwona.mironczuk-chodakowska@umb.edu.pl (I.M.-C.)

2 Department of Epidemiology, Cardiovascular Disease Prevention and Health Promotion, National Institute of Cardiology, Alpejska 42, 04-628 Warsaw, Poland; awaskiewicz@ikard.pl (A.W.); acicha@ikard.pl (A.C.-M.); wdrygas@ikard.pl (W.D.)

3 Department of Social and Preventive Medicine, Faculty of Health Sciences, Medical University of Lodz, Hallera 1, 90-001 Lodz, Poland

* Correspondence: anna.witkowska@umb.edu.pl; Tel.: +48-85-6865090; Fax: +48-85-6865089

† These authors contributed equally to this work.

Abstract: Dietary cholesterol has been suggested to increase the risk of cardiovascular disease (CVD). Phytosterols, present in food or phytosterol-enriched products, can reduce cholesterol available for absorption. The present study aimed to investigate the association between habitual intake of total and individual plant sterols (β-sitosterol, campesterol, and stigmasterol) or a diet combined with phytosterol-enriched products and CVD in a cross-section of Polish adults, participants of the Multicenter National Health Survey II (WOBASZ II). Among men (n = 2554), median intakes of plant sterols in terciles ranged between 183–456 mg/d and among women (n = 3136), 146–350 mg/d in terciles. The intake of phytosterols, when consumed with food containing phytosterols, including margarine, ranged between 184–459 mg/d for men and 147–352 mg/d for women. Among both men and women, beta-sitosterol intake predominated. Plant sterol intake was lower among both men and women with CVD (p = 0.016) compared to those without CVD. Diet quality, as measured by the Healthy Diet Index (HDI), was significantly higher in the third tercile of plant sterol intake for both men and women and the entire study group (p < 0.0001). This study suggests that habitual dietary intake of plant sterols may be associated with a lower chance of developing CVD, particularly in men.

Keywords: cardiovascular diseases; diabetes mellitus; humans; adult; phytosterols; diet; margarine

Citation: Witkowska, A.M.; Waśkiewicz, A.; Zujko, M.E.; Cicha-Mikołajczyk, A.; Mirończuk-Chodakowska, I.; Drygas, W. Dietary Plant Sterols and Phytosterol-Enriched Margarines and Their Relationship with Cardiovascular Disease among Polish Men and Women: The WOBASZ II Cross-Sectional Study. *Nutrients* **2022**, *14*, 2665. https://doi.org/10.3390/nu14132665

Academic Editor: Arrigo Cicero

Received: 3 June 2022
Accepted: 24 June 2022
Published: 27 June 2022

Publisher's Note: MDPI stays neutral with regard to jurisdictional claims in published maps and institutional affiliations.

Copyright: © 2022 by the authors. Licensee MDPI, Basel, Switzerland. This article is an open access article distributed under the terms and conditions of the Creative Commons Attribution (CC BY) license (https://creativecommons.org/licenses/by/4.0/).

1. Introduction

Cardiovascular disease (CVD) is a global health problem and a leading cause of death [1]. CVD risk factors are associated with poor lifestyle, including smoking, physical inactivity, obesity, unhealthy diet, and excessive alcohol consumption, leading to hypertension, hyperglycemia, and high LDL cholesterol [2,3]. Studies indicate a link between CVD and diabetes [3,4].

Type 2 diabetes mellitus (T2DM) predisposes patients to cardiovascular disease and cardiovascular mortality [5]. The development and progression of T2DM are strongly influenced by diet, physical inactivity, and increased body weight; therefore, intensive lifestyle modification is recommended for T2DM [6]. In patients with diabetes, the addition of soluble dietary fiber and phytosterols is recommended as a primary measure to prevent CVD before considering non-statin therapy [7].

Phytosterols (plant sterols and plant stanols) are natural bioactive plant substances with a structure similar to cholesterol. In the intestine, phytosterols and cholesterol compete for the same absorption mechanisms [8]. As a result, phytosterols can affect blood cholesterol concentrations by reducing the amount of cholesterol available for absorption. Studies have shown that consumption of 0.6–3.3 g of plant sterols per day reduces serum LDL-C concentrations by approximately 6–12%, and this effect was dose-dependent [9].

The diet typically provides 150–400 mg of plant sterols [10–15]. The phytosterols found in the highest amounts in plant-based foods, and, thus, in the human diet, are β-sitosterol, campesterol, and stigmasterol [16]. Food sources with the highest plant sterol content are vegetable oils, mainly corn oil, and sesame seeds [17]. Phytosterols isolated mainly from vegetable oils and their commercially produced esters can be ingredients of fortified foods and supplements as a non-pharmacological therapy of hypercholesterolemia. In European Union countries, products enriched in plant sterols are mainly milk and yogurt, margarine, and spreadable fats [18]. Plant sterol-enriched foods that provide 2 mg of phytosterols daily, combined with a healthy lifestyle, in patients with mild to moderate hypercholesterolemia have been found to reduce LDL-C levels by 10% [19,20]. However, the effect of long-term use of phytosterol-enriched foods on cardiovascular risk factors is unknown [21].

A few population-based studies, but not in the Polish population, have analyzed the effects of dietary phytosterol intake on CVD [10,11,14], but none included phytosterol-enriched products. Therefore, the present study aimed to investigate whether there is an association between habitual intake of total phytosterols and individual phytosterols (β-sitosterol, campesterol, and stigmasterol), or a diet combined with phytosterol-enriched products, and CVD in a cross-section of Polish adults.

2. Materials and Methods

2.1. Study Group

The study group consisted of 2554 men and 3136 women, of the National Multicenter Health Survey II (in Polish—WOBASZ II). WOBASZ II is a cross-sectional study representative of the Polish population of adults aged 20 years and older, which was conducted by the Institute of Cardiology (at present National Institute of Cardiology), Warsaw, Poland, in 2013–2014, in collaboration with five national medical universities. The design and methods of the WOBASZ II study have been described in detail elsewhere [22]. Approval for the WOBASZ II study was obtained from the Bioethics Committee at the National Institute of Cardiology (No. 1344), and was approved for the current study (No. 1837). Written informed consent was obtained from all participants.

Data on participants' demographics, diseases, leisure-time physical activities, tobacco use, and alcohol intake, were collected using a standardized questionnaire developed for the WOBASZ II study. The classification of cardiovascular disease (CVD) was adopted according to World Health Organization guidelines [23]. Respondents were defined as having CVD if they had a reported history of any of the following: coronary heart disease, myocardial infarction, stroke, atrial fibrillation and/or other cardiac arrhythmias, peripheral vascular disease of the lower limbs, heart failure, coronary angioplasty or coronary artery bypass grafting, and implanted pacemaker or cardioverter-defibrillator. The criterion for diabetes, according to the American Diabetes Association [24], was a glucose level $\geq$ 7.0 mmol/L and/or use of glucose-lowering medication. Blood pressure (BP) was measured three times on the right arm after 5 min of rest in a sitting position at 1 min intervals, and the final BP was reported as the mean of the second and third measurements. Hypertension was diagnosed when systolic blood pressure was $\geq$140 mmHg and/or diastolic blood pressure $\geq$ 90 mmHg and/or when antihypertensive drugs were used. Height and weight measurements were taken by personnel trained in standard procedures. Body mass index (BMI) was calculated from body weight in kilograms divided by the square of height in meters. Biochemical analyses were performed at the Central Laboratory "Diagnostyka" at the National Institute of Cardiology in Warsaw.

2.2. Food Intake and Nutritional Assessment

Data on daily food intake were collected by trained interviewers using the single 24-h dietary recall method. To reduce the possibility of bias, individuals who described their diet as atypical were excluded. Based on the different types of food consumed, energy and dietary fiber of each patient's diet were calculated using Polish food composition tables [25]. Polyphenols and antioxidants were calculated using previous studies [26–29].

2.3. Assessment of Healthy Diet Index (HDI) Score

Diet quality was determined by scoring the Healthy Diet Indicator (HDI), which was in accordance with the World Health Organization (WHO) dietary guidelines [18] and described in Fransen et al. [30]. HDI is based on six components—intake of saturated fatty acids (% total energy, %TE), intake of polyunsaturated fatty acids (%TE), dietary cholesterol (mg/d), dietary protein (%TE), fiber (g/d), and free sugars (%TE)—and fruits and vegetables (g/d), within the recommended range [31]. The final HDI score was the sum of all components, ranging from zero (minimal compliance with recommendations) to seven (maximum compliance with recommendations).

2.4. Assessment of Dietary Phytosterol Intake and the Intake of Plant Sterols from Enriched Margarine

Phytosterol intake was calculated as previously described using a developed database [12]. Total and individual phytosterol intakes were determined by multiplying the daily intake of each food by the total and individual phytosterol content of that food, respectively. Dietary recalls were reviewed by checking for consumption of phytosterol-enriched products. Based on the dietary history it was found that among the products enriched with phytosterols, only phytosterol-enriched margarine was consumed by 1.96% of men and 1.85% of women [12]. Manufacturers were identified and asked to report the plant sterol content of their products.

2.5. Statistical Analysis

The study population was divided into three groups according to the tercile distributions of plant sterol intakes (separately for total and individual phytosterols). All analyses were performed according to gender and overall. Quantitative variables were presented as mean (standard deviation) and/or median (interquartile range), while qualitative variables were presented as percentages. Mean values of plant sterol intake with a 95% confidence interval (95% CI), adjusted for age, were calculated using the general linear model and the Tukey-Kramer test was chosen for multiple comparisons, if appropriate. The odds ratios (ORs) with 95% CI for CVD were evaluated using logistic regression analysis in relation to total and particular phytosterol intake. Two models were applied: model 1, unadjusted in men and women but adjusted for sex, and combined, and model 2, adjusted for age, consumption of lipid-lowering drugs, HDI, BMI, alcohol intake, and, additionally, for sex, for the entire population. The first tercile (T1) in each model was adopted as a reference. Statistical analyses were carried out using SAS software version 9.4 (SAS Institute Inc., Cary, NC, USA). A p-value less than 0.05 was considered statistically significant.

3. Results

The general characteristics of the study participants are shown in Table 1. The mean age of the entire study group was 49.58 years. The highest percentage of the study participants had hypercholesterolemia 67.3% and hypertension 45.22%. CVD was diagnosed in 20% of the studied population, while diabetes in 10.82%.

Table 2 shows the phytosterol intake according to age, presence of diabetes, and CVD. The results are presented for men, women, and the entire study group. Dietary phytosterol content was found to be age-dependent and generally highest among the youngest age group and lowest among those aged 65 years and older. Among men and in the entire study group, sterol intake was significantly lower among people with diabetes (results were adjusted for age). No significant differences were found for women. With respect to CVD, plant sterol intake was lower among both men and women with CVD ($p = 0.0016$) and for both genders ($p < 0.0001$). With regard to diabetes, such a relationship was observed for men and the entire group, but not for women. With respect to individual plant sterols, we found that dietary intake of phytosterols was lower among both men with CVD and women and among men with diabetes (except campesterol in men with diabetes). No differences were found between women with diabetes and healthy women. For the whole group, only campesterol was not statistically significant. The intake of individual plant sterols with the fortified margarine was not considered, because manufacturers only reported the total phytosterol content. Thus, it was not possible to determine what the individual plant sterol content of the margarine was.

Table 3 shows terciles of plant sterol intake with food, and with food including phytosterol-enriched margarine. Terciles of individual plant sterol intake for the entire study group and by gender were used as means (crude, adjusted), medians, and ranges for particular phytosterols intake. Among men, the median plant sterol intake in the first tercile was 183, in the second tercile 292, and in the third tercile 456 mg/d. For food intake, including margarine with phytosterols, the values were 184; 294, and 459 mg/d, respectively. Among women, the median intakes of plant sterols with diet were: 146 in the first tercile, 231 in the second tercile, and 350 mg/d in the third tercile. For food intake, including margarine with phytosterols, these values for women were, respectively: 147; 232, and 352 mg/d. For individual plant sterols, they are ranked in Table 3 by the volume of intake. Among both men and women, beta-sitosterol intake predominated, with a median range of 112–280 mg/d per tercile among men and 91–222 mg/d among women. For campesterol, the median range was 31–107 mg/d among men and 24–78 mg/d among women, and for stigmasterol, 12–39 mg/d among men and 12–34 mg/d for women.

The odds ratio of developing CVD was related to phytosterol intake with diet (Figure 1). In the crude model, it was found that in both men and women, and in the entire study group (adjusted for gender), OR of CVD were significantly lower in the second and third terciles compared to the first terciles, with the lowest incidence of CVD in the third tercile. After adjusting for confounding factors, among men statistical significance was maintained, except for the second tercile of beta-sitosterol intake. Among women, only the intake of total plant sterols from the diet and their total intake together with margarine in the third tercile, and the intake of beta-sitosterol in the third tercile remained statistically significant. In the entire study group, significant values were observed in the third tercile of total plant sterol intake (without and with phytosterol-enriched products), and for all individual plant sterols.

Table 1. General characteristics of the studied population.

Trait	Men $n = 2554$	Women $n = 3136$	Total $n = 5690$
Age (year), mean ± SD	48.79 ± 16.27	50.23 ± 16.54	49.58 ± 16.43
BMI (kg/m^2), mean ± SD	27.42 ± 4.55	26.96 ± 5.65	27.17 ± 5.19
CVD, (%)	19.34	20.54	20.00
Hypertension, (%) [1]	49.56	41.69	45.22
Hypercholesterolemia, (%) [2]	68.86	66.03	67.30
Diabetes, (%) [3]	11.86	9.96	10.82

[1] Hypertension: systolic blood pressure SBP ≥ 140 mmHg or diastolic blood pressure DBP ≥ 90 mmHg, or use of antihypertensive drugs. [2] Hypercholesterolemia: total cholesterol ≥ 5 mmol/L or LDL cholesterol ≥ 3 mmol/L or the participant was taking lipid-lowering medication. [3] Diabetes: blood glucose level was ≥7.0 mmol/L or diabetes was declared in an interview.

Table 2. Phytosterol intakes depending on age, diabetes, and CVD.

	Men		**Women**		**Total**	
	n **= 2554**		***n*** **= 3136**		***n*** **= 5690**	
	Mean (95% CI)	***p***	**Mean (95% CI)**	***p***	**Mean (95% CI)**	***p***
Total plant sterol intake from food (mg/d)						
Age (years)						
20–44	344.3 (333.0–351.9)	<0.0001	258.5 (251.7–265.2)	<0.0001	300.0 (294.3–305.6)	<0.0001
45–64	323.0 (313.1–332.8)		263.1 (256.5–269.7)		293.5 (287.8–299.2)	
65+	264.1 (249.7–278.4)		217.8 (208.5–227.1)		242.5 (234.3–250.7)	
Diabetes (mg/d)						
no	323.3 (316.6–330.1)	0.0379	253.6 (249.0–258.2)	0.1289	288.4 (284.4–292.4)	0.0077
yes	301.7 (282.6–320.8)		241.9 (227.5–256.2)		271.3 (259.6–283.1)	
CVD (mg/d)						
no	326.1 (319.1–333.1)	0.0016	255.9 (251.1–260.7)	0.0016	290.9 (286.8–295.0)	<0.0001
yes	298.6 (283.5–313.6)		237.8 (227.9–247.7)		268.3 (259.7–277.0)	
Total phytosterols intake from food and enriched margarine (mg/d)						
Age (years)						
20–44	344.3 (334.7–353.9)	<0.0001	258.9 (252.1–265.7)	<0.0001	301.1 (295.4–306.9)	<0.0001
45–64	326.0 (316.0–336.0)		264.9 (258.3–271.6)		295.9 (290.1–301.7)	
65+	266.6 (252.0–281.2)		219.6 (210.2–229.0)		244.8 (236.4–253.1)	
Diabetes (mg/d)						
no	325.3 (318.4–332.2)	0.0789	255.1 (250.4–259.7)	0.0941	290.1 (286.1–294.2)	0.0121
yes	306.6 (287.2–326.1)		242.0 (227.5–256.4)		273.9 (261.9–285.8)	
CVD (mg/d)						
no	328.2 (321.1–335.3)	0.0034	257.2 (252.4–262.1)	0.0016	292.7 (288.5–296.9)	<0.0001
yes	302.3 (286.9–317.6)		238.9 (228.9–248.9)		270.6 (261.8–279.4)	
Beta-sitosterol (mg/d)						
Age (years)						
20–44	209.7 (204.0–215.5)	<0.0001	160.9 (156.8–165.0)	<0.0001	185.0 (181.5–188.5)	<0.0001
45–64	200.1 (194.1–206.2)		165.9 (161.9–170.0)		183.3 (179.8–186.8)	
65+	164.5 (155.7–173.3)		137.0 (131.3–142.6)		151.6 (146.6–156.6)	
Diabetes (mg/d)						
no	199.3 (195.2–203.4)	0.0468	159.1 (156.3–161.9)	0.0840	179.2 (176.7–181.6)	0.0065
yes	186.6 (175.0–198.3)		150.9 (142.2–159.7)		168.5 (161.3–175.7)	
CVD (mg/d)						
no	200.7 (196.4–205.0)	0.0062	160.4 (157.5–163.3)	0.0016	180.5 (178.0–183.1)	<0.0001
yes	186.1 (176.9–195.3)		149.4 (143.3–155.4)		167.7 (162.4–173.0)	
Campesterol						
Age (years)						
20–44	75.6 (72.8–78.4)	<0.0001	52.6 (50.7–54.5)	<0.0001	63.9 (62.3–65.6)	<0.0001
45–64	68.4 (65.5–71.4)		52.9 (51.0–54.7)		60.8 (59.1–62.5)	
65+	55.3 (51.0–59.6)		43.3 (40.7–45.9)		49.7 (47.4–52.1)	
Diabetes (mg/d)						
no	69.5 (67.5–71.5)	0.3479	50.9 (49.7–52.2)	0.6694	60.2 (59.1–61.4)	0.2715
yes	66.6 (60.9–72.3)		50.0 (46.1–54.0)		58.2 (54.8–61.6)	
CVD (mg/d)						
no	70.6 (68.5–72.6)	0.0044	51.9 (50.5–53.2)	0.0015	61.2 (60.0–62.4)	<0.0001
yes	63.2 (58.7–67.7)		46.9 (44.1–49.6)		55.1 (52.6–57.5)	
Stigmasterol						
Age (years)						
20–44	28.0 (27.0–28.9)	<0.0001	24.1 (23.4–24.8)	<0.0001	26.0 (25.4–26.6)	<0.0001
45–64	27.0 (26.0–28.0)		24.5 (23.8–25.2)		25.8 (25.2–26.3)	
65+	21.2 (19.8–22.7)		19.0 (18.0–20.0)		20.2 (19.4–21.0)	
Diabetes (mg/d)						
no	26.8 (26.2–27.5)	0.0007	23.4 (22.9–23.9)	0.1717	25.1 (24.7–25.5)	0.0004
yes	23.3 (21.4–25.2)		22.3 (20.8–23.8)		22.8 (21.6–24.0)	
CVD (mg/d)						
no	26.9 (26.2–27.6)	0.0017	23.6 (23.1–24.1)	0.0023	25.2 (24.8–25.7)	<0.0001
yes	24.2 (22.6–25.7)		21.7 (20.7–22.8)		23.0 (22.1–23.9)	

Results adjusted for age in men and women and additionally for sex in total; adjustment not applicable to the age groups.

Table 3. Intake of phytosterols in terciles.

	Men			Women			Total		
	N = 2554			N = 3136			N = 5690		
	Tercile 1	**Tercile 2**	**Tercile 3**	**Tercile 1**	**Tercile 2**	**Tercile 3**	**Tercile 1**	**Tercile 2**	**Tercile 3**
	n = 851	*n* = 851	*n* = 852	*n* = 1045	*n* = 1045	*n* = 1046	*n*=1896	*n*=1897	*n*=1897
Total phytosterol intake from food (mg/d)									
Mean ± SD Me (IQR) Range	173.8 ± 43.9 183.2 (142.0–210.0) 0.23–234.7	292.3 ± 32.8 291.6 (265.1–319.6) 234.7–353.6	496.0 ± 146.9 455.9 (398.7–547.1) 354.0–1774.0	140.6 ± 33.6 146.4 (119.2–167.7) 27.2–190.2	233.1 ± 26.1 230.6 (210.0–254.4) 190.3–281.7	382.7 ± 111.4 349.9 (308.1–418.5) 281.8–1632.7	152.8 ± 38.1 158.0 (127.0–185.0) 0.23–207.8	257.2 ± 29.9 255.9 (229.8–283.4) 207.8–310.6	438.8 ± 135.7 399.8 (348.0–487.9) 310.7–1774.0
Adjusted mean (95% CI) *	174.8 (168.7–180.9)	292.5 (286.5–298.6)	494.8 (488.7–500.9)	140.9 (136.7–145.1)	233.1 (228.9–237.2)	382.5 (378.3–386.6)	156.6 (152.8–160.4)	258.4 (254.6–262.1)	436.4 (432.7–440.2)
Total phytosterol intake from food and enriched margarine (mg/d)									
Mean ± SD Me (IQR) Range	174.2 ± 44.0 183.9 (142.5–210.4) 0.23–235.2	294.0 ± 33.3 294.0 (266.7–321.0) 235.3–356.2	501.1 ± 150.9 459.1 (403.8–550.6) 356.4–1774.0	141.0 ± 33.9 146.9 (119.2–168.4) 27.2–191.2	234.2 ± 26.2 232.2 (211.3–255.5) 191.2–283.1	385.1 ± 112.1 352.0 (309.6–422.5) 283.1–1632.7	153.2 ± 38.3 158.9 (127.1–185.4) 0.23–208.6	258.4 ± 30.1 257.2 (230.9–284.6) 208.8–312.1	442.6 ± 138.4 404.5 (350.3–491.1) 312.1–1774.0
Adjusted mean (95% CI) *	175.2 (168.9–181.4)	294.3 (288.0–300.5)	500.0 (493.7–506.3)	141.2 (137.0–145.4)	234.2 (230.0–238.4)	384.9 (380.7–389.1)	157.0 (153.1–160.8)	259.6 (255.8–263.4)	440.2 (436.4–444.1)
B-sitosterol (mg/d)									
Mean ±SD Me (IQR) Range	107.3 ± 27.9 112.4 (87.1–131.2) 0.20–146.0	181.1 ± 20.3 180.8 (163.9–197.8) 146.0–219.3	305.2 ± 88.1 279.6 (246.9–332.9) 219.3- 1028.7	87.6 ± 22.0 90.5 (72.8–105.9) 14.8–120.0	147.0 ± 16.3 146.2 (133.0–161.4) 120.0–175.9	239.7 ± 62.7 222.0 (196.1–264.3) 176.0–679.6	95.0 ± 24.6 98.35 (77.4–115.8) 0.20–130.7	160.9 ± 18.3 160.8 (144.4–176.6) 130.8–194.6	272.0 ± 79.3 248.2 (218.1–301.3) 194.7–1028.7
Adjusted mean (95% CI) *	107.8 (104.1–111.4)	181.2 (177.5–184.8)	304.6 (300.9–308.3)	87.8 (85.4–90.2)	146.9 (144.5–149.3)	239.6 (237.2–242.0)	97.2 (95.0–99.4)	161.5 (159.3–163.7)	270.8 (268.6–273.0)
Campesterol (mg/d)									
Mean ±SD Me (IQR) Range	30.2 ± 8.6 31.4 (23.9–37.2) 0.02–43.5	57.2 ± 8.5 56.7 (49.7–64.1) 43.5–73.8	120.0 ± 49.3 106.6 (86.4–137.8) 73.9–585.8	23.0 ± 6.0 23.7 (18.7–27.8) 3.2- 32.2	41.7 ± 6.1 40.9 (36.4–46.4) 32.2–54.2	87.8 ± 32.3 78.1 (64.9–100.8) 54.3–342.3	25.5 ± 7.1 26.1 (20.6–31.5) 0.02–36.3	48.2 ± 7.8 47.4 (41.4–54.4) 36.3–63.5	103.4 ± 42.8 90.7 (74.4–118.5) 63.5–585.8
Adjusted mean (95% CI) *	30.6 (28.6–32.6)	57.2 (55.2–59.1)	119.7 (117.7–121.6)	23.0 (21.8–24.2)	41.7 (40.6–42.9)	87.7 (86.6–88.9)	26.7 (25.5–27.9)	48.5 (47.3–49.6)	102.9 (101.7–104.0)
Stigmasterol (mg/d)									
Mean ±SD Me (IQR) Range	11.9 ± 3.9 12.2 (9.1–15.1) 0.003–18.3	23.6 ± 3.2 23.5 (20.7–26.4) 18.3–29.2	43.6 ± 15.3 39.1 (32.9–48.1) 29.3–131.7	11.2 ± 3.5 11.7 (8.7–14.2) 0.9–16.4	21.2 ± 2.7 21.1 (18.8–23.6) 16.4–26.0	37.3 ± 11.6 33.9 (29.2–41.8) 26.1–132.3	11.5 ± 3.7 11.8 (8.9–14.5) 0.003–17.2	22.2 ± 2.9 22.1 (19.6–24.7) 17.2–27.3	40.2 ± 13.6 36.1 (30.9–44.9) 27.4–132.3
Adjusted mean (95% CI) *	12.0 (11.4–12.6)	23.6 (23.0–24.2)	43.5 (42.8–44.1)	11.3 (10.8–11.7)	21.1 (20.7–21.6)	37.3 (36.8–37.7)	11.6 (11.3–12.0)	22.3 (21.9–22.6)	40.1 (39.8–40.5)

* Results were adjusted for age in men and women and additionally for sex in total.

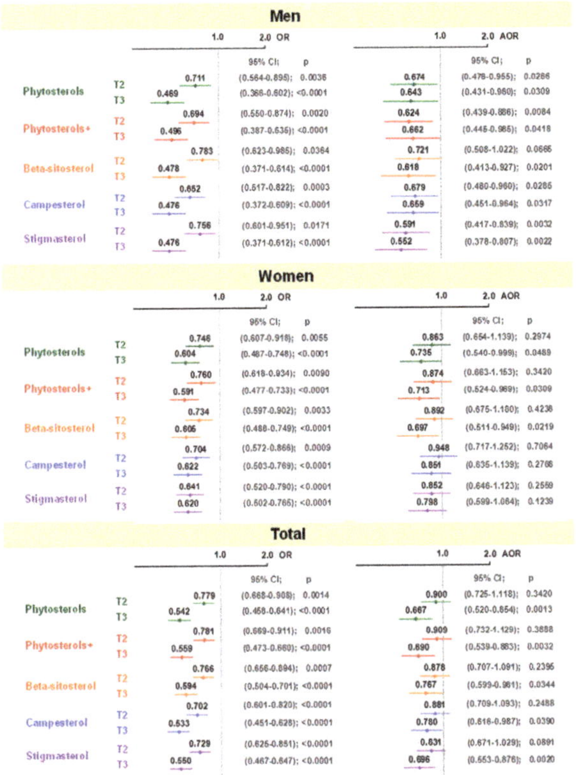

Figure 1. Odds ratio (95% confidence interval) for CVD in relation to total and individual phytosterol intake (relative to 1st tercile). T2—2nd tercile; T3—3rd tercile; OR—odds ratio; AOR—adjusted odds ratio; ORs were unadjusted in men and women but adjusted for sex combined; AORs—adjusted for age, lipid-lowering medication, HDI, BMI, alcohol, and additionally for sex in total.

We also investigated whether the results obtained could be biased by diet (Table 4). For this purpose, data from the extreme terciles of total and single plant sterol intake before and after energy adjustment were presented by sex and the entire study group. It was found that both before and after adjustment the results were significant for total and single dietary plant sterol. For intakes of phytosterol-enriched margarine, a significant difference was found between the first and third terciles before energy adjustment, which did not occur after adjustment. Intakes of polyphenols, antioxidants, dietary fiber, and HDI were also divided according to the tercile of total and individual plant sterol intake, with polyphenols, antioxidants, and dietary fiber adjusted for energy value. It was found that before adjustment, dietary polyphenol, antioxidant, and fiber contents were higher in the third tercile among both men and women and in the group as a whole ($p < 0.0001$). After adjustment for energy, differences were not observed. Diet quality, as measured by HDI, was significantly higher in the third tercile of plant sterol intake for both men and women and for the entire study group ($p < 0.0001$).

Intakes of atherogenic and antiatherogenic products were also examined in the first and third terciles of total and individual phytosterol intake (Table 5). For atherogenic products, butter and animal fat consumption was found to be higher in the third tercile of plant sterol intake, but after adjustment for energy there was an inverse difference, i.e., with higher plant sterol intake, animal fat and butter consumption was lower. For red meat and meat products before and after adjustment for energy, consumption was higher in the third tercile. All the above observations were true for both men and women and for the entire study group.

Table 4. Dietary quality in relation to tercile of dietary total and individual phytosterol intake (1st tercile vs. 3rd tercile).

	Men			Women			Total		
	N = 2554			N = 3136			N = 5960		
	1st Tercile	3rd Tercile		1st Tercile	3rd Tercile		1st Tercile	3rd Tercile	
	n = 851	n = 852	p	n = 1045	n = 1046	p	n = 1896	n = 1897	p
	Mean (95% CI)	Mean (95% CI)		Mean (95% CI)	Mean (95% CI)		Mean (95% CI)	Mean (95% CI)	
Phytosterols									
Total plant sterols (mg/d)	174.8 (168.7–180.9)	494.8 (488.7–500.9)	<0.0001	140.9 (136.7–145.1)	382.5 (378.3–386.6)	<0.0001	156.6 (152.8–160.4)	436.4 (432.7–440.2)	<0.0001
Total plant sterols/1000 kcal (mg/d)	111.6 (108.8–114.4)	171.8 (169.1–174.6)	<0.0001	124.1 (121.2–127.1)	189.7 (186.7–192.6)	<0.0001	117.9 (115.9–120.0)	180.9 (178.8–183.0)	<0.0001
B-sitosterol (mg/d)	108.4 (104.6–112.1)	303.9 (300.1–307.7)	<0.0001	88.6 (86.1–91.1)	238.5 (236.1–241.0)	<0.0001	97.6 (95.3–99.9)	269.7 (267.4–271.9)	<0.0001
B-sitosterol/1000 kcal (mg/d)	69.2 (67.5–71.0)	105.7 (103.9–107.4)	<0.0001	78.1 (76.2–79.9)	118.2 (116.4–120.1)	<0.0001	72.8 (71.4–74.1)	111.8 (110.5–113.1)	<0.0001
Campesterol (mg/d)	32.6 (30.4–34.8)	115.5 (113.3–117.7)	<0.0001	24.5 (23.1–25.8)	84.2 (82.8–85.5)	<0.0001	28.2 (26.9–29.5)	98.9 (97.6–100.1)	<0.0001
Campesterol/1000 kcal (mg/d)	20.7 (19.8–21.6)	40.1 (39.2–41.0)	<0.0001	21.4 (20.6–22.3)	42.1 (41.2–42.9)	<0.0001	20.5 (19.9–21.1)	41.0 (40.3–41.6)	<0.0001
Stigmasterol (mg/d)	16.2 (15.3–17.1)	37.6 (36.7–38.5)	<0.0001	14.9 (14.2–15.5)	31.4 (30.8–32.1)	<0.0001	15.6 (15.1–16.2)	34.4 (33.9–35.0)	<0.0001
Stigmasterol/1000 kcal (mg/d)	10.7 (10.3–11.2)	13.2 (12.7–13.7)	<0.0001	13.5 (13.0–14.1)	15.6 (15.1–16.1)	<0.0001	12.2 (11.9–12.6)	14.4 (14.0–14.7)	<0.0001
Phytosterols from enriched margarine (mg/d)	1.03 (0.00–2.52)	3.81 (2.32–5.30)	0.0278	0.66 (0.02–1.31)	2.05 (1.41–2.70)	0.0083	0.80 (0.03–1.57)	2.93 (2.17–3.69)	0.0004
Phytosterols from enriched margarine/1000 kcal (mg/d)	0.69 (0.03–1.35)	1.42 (0.76–2.09)	0.2832	0.59 (0.20–0.99)	1.09 (0.70–1.49)	0.1861	0.57 (0.19–0.95)	1.25 (0.88–1.62)	0.0330
Polyphenols and antioxidants									
Polyphenols (mg/d)	1464 (1410–1518)	2705 (2651–2759)	<0.0001	1427 (1383–1472)	2494 (2450–2538)	<0.0001	1443 (1408–1478)	2594 (2559–2629)	<0.0001
Polyphenols/1000 kcal (mg/d)	945 (916–973)	945 (916–973)	0.9999	1260 (1224–1296)	1227 (1191–1263)	0.4125	1112 (1088–1136)	1086 (1062–1110)	0.2963
Antioxidants (mmol/d)	8.71 (8.25–9.18)	16.20 (15.74–16.67)	<0.0001	8.84 (8.43–9.25)	15.36 (14.95–15.77)	<0.0001	8.79 (8.48–9.11)	15.78 (15.47–16.09)	<0.0001
Antioxidants/1000 kcal (mmol/d)	5.70 (5.47–5.93)	5.71 (5.48–5.94)	0.9979	7.87 (7.57–8.17)	7.60 (7.29–7.90)	0.4324	6.86 (6.66–7.07)	6.66 (6.46–6.86)	0.3449
Dietary fiber									
Dietary fiber (g/d)	14.9 (14.4–15.4)	27.5 (27.0–28.0)	<0.0001	12.4 (12.0–12.8)	22.5 (22.1–22.9)	<0.0001	13.5 (13.2–13.9)	24.9 (24.6–25.2)	<0.0001
Dietary fiber/1000 kcal (g/d)	9.4 (9.1–9.6)	9.6 (9.4–9.8)	0.3899	10.7 (10.4–11.0)	11.1 (10.8–11.4)	0.1432	10.0 (9.9–10.2)	10.3 (10.1–10.5)	0.1132
Diet quality									
HDI score (points)	2.67 (2.59–2.75)	3.83 (3.75–3.92)	<0.0001	2.76 (2.69–2.83)	3.85 (3.78–3.91)	<0.0001	2.70 (2.65–2.76)	3.83 (3.77–3.88)	<0.0001

Results were adjusted for age in men and women and additionally for gender in men and women overall.

Table 5. Consumption of selected products by tercile of phytosterol intake (1st tercile vs. 3rd tercile).

	Men			Women			Total		
	n = 2554			n = 3136			n = 5960		
	1st Tercile	3rd Tercile		1st Tercile	3rd Tercile		1st Tercile	3rd Tercile	
	n = 851	n = 852	p	n = 1045	n = 1046	p	n = 1896	n = 1897	p
	Mean (95% CI)	Mean (95% CI)		Mean (95% CI)	Mean (95% CI)		Mean (95% CI)	Mean (95% CI)	
Atherogenic food									
Butter (g/d)	13.1 (11.7–14.5)	16.9 (15.5–18.3)	0.0007	10.3 (9.4–11.2)	12.7 (11.8–13.6)	0.0005	11.8 (11.0–12.6)	14.9 (14.1–15.7)	<0.0001
Butter/1000 kcal (g/d)	7.8 (7.3–8.4)	5.3 (4.8–5.8)	<0.0001	8.2 (7.7–8.7)	5.8 (5.3–6.4)	<0.0001	8.1 (7.7–8.5)	5.6 (5.2–6.0)	<0.0001
Red meat and cold cuts (g/d)	103.3 (97.8–118.9)	214.2 (203.7–224.8)	<0.0001	53.6 (48.2–58.9)	101.9 (96.5–107.3)	<0.0001	84.1 (78.3–89.8)	156.4 (150.7–162.0)	<0.0001
Red meat and cold cuts/1000 kcal (g/d)	61.4(57.4–65.3)	69.7(65.8–73.6)	0.0102	41.0 (38.0–44.1)	48.1 (45.1–51.2)	0.0034	51.7 (49.2–54.2)	58.7 (56.3–61.2)	0.0003
Animal fats (g/d)	23.2 (21.2–25.3)	28.2 (26.1–30.2)	0.0024	18.6 (17.3–19.8)	20.4 (19.1–21.6)	0.1120	21.1 (19.9–22.2)	24.5 (23.4–25.7)	0.0001
Animal fats/1000 kcal (g/d)	13.4 (12.7–14.1)	8.7 (8.0–9.4)	<0.0001	14.4 (13.8–15.1)	9.1 (8.5–9.8)	<0.0001	14.0 (13.6–14.5)	9.0 (8.5–9.5)	<0.0001
Antiatherogenic food									
Oils (g/d)	2.3 (1.8–3.9)	25.3 (24.3–26.4)	<0.0001	1.9 (1.2–2.6)	19.4 (18.8–20.1)	<0.0001	2.2 (1.5–2.8)	22.0 (21.4–22.6)	<0.0001
Oils/1000 kcal (g/d)	1.9 (1.5–2.4)	8.9 (8.5–9.3)	<0.0001	1.6 (1.3–2.0)	10.0 (9.6–10.4)	<0.0001	1.5 (1.2–1.8)	9.3 (9.0–9.6)	<0.0001
Soft margarine (g/d)	6.9 (5.8–8.1)	18.3 (17.1–19.4)	<0.0001	4.7 (4.1–5.4)	9.8 (9.1–10.4)	<0.0001	6.0 (5.4–6.7)	13.9 (13.3–14.6)	<0.0001
Soft margarine/1000 kcal (g/d)	4.5 (4.0–5.0)	6.4 (5.9–6.9)	<0.0001	4.4 (4.0–4.8)	4.8 (4.4–5.2)	0.3933	4.5 (4.2–4.8)	5.6 (5.3–5.9)	<0.0001
Vegetable fats (g/d)	9.7 (8.3–11.2)	43.7 (42.3–45.1)	<0.0001	6.6 (5.8–7.5)	29.3 (28.4–30.1)	<0.0001	8.2 (7.3–9.0)	36.0 (35.2–36.8)	<0.0001
Vegetable fats/1000 kcal (g/d)	6.4 (5.8–7.0)	15.3 (14.7–15.9)	<0.0001	6.0 (5.5–6.6)	14.8 (14.3–15.3)	<0.0001	6.0 (5.6–6.4)	15.0 (14.6–15.4)	<0.0001
Fish (g/d)	11.7(7.1–16.3)	37.3(32.7–41.9)	<0.0001	7.6 (4.4–10.7)	25.7 (22.5–28.9)	<0.0001	8.8 (6.0–11.6)	31.3 (28.6–34.0)	<0.0001
Fish/1000 kcal (g/d)	7.5 (5.5–9.5)	13.3 (11.3–15.4)	0.0002	6.5 (4.5–8.4)	13.2 (11.2–15.2)	<0.0001	6.4 (4.9–7.8)	13.4 (11.9–14.8)	<0.0001
Wholemeal bread (g/d)	22.0 (17.6–26.5)	34.0 (29.5–38.5)	0.0007	18.5 (15.5–21.6)	32.4 (29.4–35.5)	<0.0001	19.2 (16.5–21.8)	32.6 (29.9–35.2)	<0.0001
Wholemeal bread/1000 kcal (g/d)	14.3 (12.1–16.4)	13.1 (10.9–15.2)	0.7170	17.8 (15.8–19.9)	16.5 (14.5–18.6)	0.6651	15.7 (14.2–17.2)	14.6 (13.1–16.1)	0.5920
Vegetables (g/d)	195 (183–208)	316 (304–328)	<0.0001	168 (159–178)	287 (277–296)	<0.0001	181 (173–188)	301 (293–309)	<0.0001
Vegetables/1000 kcal (g/d)	131 (124–138)	113 (106–120)	0.0013	153 (145–160)	146 (139–154)	0.4727	142 (137–147)	129 (124–135)	0.0024
Fruit (g/d)	119 (103–135)	268 (253–284)	<0.0001	134 (120–147)	303 (290–317)	<0.0001	123 (112–133)	282 (272–292)	<0.0001
Fruit/1000 kcal (g/d)	77 (69–84)	95 (87–102)	0.0018	119 (111–128)	149 (141–158)	<0.0001	98 (92–104)	121 (115–126)	<0.0001
Legume (g/d)	0.73(0.00–1.88)	7.73 (6.58–8.88)	<0.0001	0.77 (0.00–1.58)	5.64 (4.83–6.46)	<0.0001	0.85 (0.16–1.55)	6.74 (6.05–7.43)	<0.0001
Legume/1000 kcal (g/d)	0.53 (0.00–1.07)	3.00 (2.46–3.54)	<0.0001	0.79 (0.23–1.35)	3.07 (2.51–3.64)	<0.0001	0.69 (0.28–1.09)	3.07 (2.68–3.47)	<0.0001
Nuts (g/d)	0.60 (0.00–1.34)	2.92 (2.18–3.67)	<0.0001	0.21 (0.00–0.92)	2.91 (2.20–3.62)	<0.0001	0.21 (0.00–0.74)	3.10 (2.58–3.62)	<0.0001
Nuts/1000 kcal (g/d)	0.35 (0.04–0.65)	1.05 (0.74–1.36)	0.0045	0.14 (0.00–0.49)	1.34 (0.99–1.69)	<0.0001	0.14 (0.00–0.38)	1.29 (1.06–1.53)	<0.0001
Seeds (g/d)	0.11 (0.00–0.63)	1.34 (0.82–1.86)	0.0033	0.12 (0.00–0.40)	1.13 (0.85–1.40)	<0.0001	0.11 (0.00–0.39)	1.20 (0.92–1.48)	<0.0001
Seeds/1000 kcal (g/d)	0.10 (0.00–0.32)	0.51 (0.29–0.73)	0.0285	0.11 (0.00–0.27)	0.57 (0.41–0.74)	0.0003	0.10 (0.00–0.24)	0.54 (0.40–0.67)	<0.0001

Results were adjusted for age in men and women and additionally for sex in total.

In the case of intake of antiatherogenic products, it was found that in both sexes and in the entire study group, both before and after adjustment for energy, the intake of vegetable oils, vegetable fats, fish, fruits, legumes, nuts, and seeds was higher in the third tercile of plant sterol intake. For soft margarine and vegetables, there were similar differences among men and the overall study group, but not among women. In women, after adjustment, differences were not observed. For whole grain bread, higher consumption by both sexes and in the entire study group was observed in the third tercile, but after adjusting for energy, differences were not significant.

4. Discussion

The prevalence of CVD and its risk factors among Poles is high [32]. CVD in this present population-based cross-sectional study was found in one fifth of the participants, which is concordant with the literature. This population requires interventions to reduce the incidence of CVD. One of the non-pharmacological treatment measures is a dietary modification to improve the quality of nutrition. Phytosterols, present in food and phytosterol-enriched food products, depending on the dose, can be effective in reducing LDL cholesterol, which is one of the risk factors for CVD [9].

Scientific evidence based on supplementation studies shows that the intake of 2 g of phytosterols is effective in lowering LDL cholesterol [20]. The relationship between dietary phytosterols and CVD is, however, controversial, as foods provide phytosterols in lower doses than dietary supplements do. The usual intake of phytosterols is generally less than 400 mg/day [10–15], and higher levels have been found only in vegans [33]. In this study, intakes higher than 400 mg/day were observed only in the highest tercile of phytosterol consumption, both in men and women. Previous evidence indicates, however, that phytosterols from natural foods may have an LDL cholesterol lowering effect [9]. In this study, both men and women with CVD were found to have lower intakes of total and individual plant sterols from diet and from diet and phytosterol-enriched margarine, than their healthy counterparts.

Diabetes predisposes one to CVD and people identified with diabetes are at a greater risk of developing cardiovascular diseases [5]. Scientific evidence shows that plant sterols can have beneficial effects on diabetes by reducing insulin resistance [34]. In this study, men with diabetes had significantly lower intakes of total and individual plant sterols, but no significant difference was observed in women.

Recent studies conducted in Poland support the belief that it is men who require special preventive measures to reduce cardiovascular risk factors, especially hypertension, dyslipidemia, diabetes, excessive body weight, and smoking [32,35]. Our cross-sectional study suggests that it is men who may benefit from habitual plant sterol intake. This is particularly evident after adjusting plant sterol intake for confounding variables, which were age, lipid-lowering medication, HDI, BMI, and alcohol. Among women, the findings are ambiguous because, after adjusting for confounders, most of the previously significant differences were not further observed for the second tercile of total and individual plant sterol intake. This might be due to the generally lower intake of plant sterols among women relative to that observed among men, and in the second quartile, it is low enough to observe beneficial effects. It is only in the third tercile of total and individual plant sterol intake that a lower incidence of CVD is observed among women.

The results of our study are in line with those of a Swedish study, which found that consumption of naturally occurring plant sterols was associated with a lower risk of a first heart attack in men, but not among women [10]. It is possible that women may benefit not from a single dietary component, but from a combination of foods and nutrients, which, for example, can be found in the Mediterranean diet or Dietary Portfolio [36]. Dietary recommendations to date regarding the consumption of a varied diet, and particularly emphasizing the consumption of plant-derived products, are reasonable in terms of providing various compounds of importance in the prevention of noncommunicable diseases. The contribution of phytosterols to the diet is highlighted by the Dietary Portfolio, which uses a combination of established nutritional approaches to lowering cholesterol, such as consumption of plant protein, nuts, soluble fiber, and monounsaturated fats and phytosterols [37,38]. It has been shown to improve LDL cholesterol fraction and other CVD risk factors [37–39]. In several other studies, lower levels of total cholesterol and LDL cholesterol were observed in relation to dietary phytosterols [40–42]. A recent study found that closer adherence to a plant-based diet was significantly associated with a lower risk of total CVD, coronary heart disease, and heart failure in postmenopausal women [36]. In contradiction to the Swedish study is the Danish study, which found no reduced CVD risk despite lower LDL-C concentrations in men [14]. However, the authors concluded that the study population had a narrow range of phytosterol intake.

Our study suggests that, in terms of intake of substances with beneficial effects on CVD, such as polyphenols, antioxidants, and dietary fiber, individuals with low and high intakes of plant sterols do not differ. However, they do differ in their intake of foods considered pro- and anti-atherogenic. It was found that study participants who had higher plant sterol intakes consumed more anti-atherogenic foods and fewer animal fats.

Phytosterol-enriched foods are recommended for people with hypercholesterolemia for the prevention of CVD [19]. However, in the WOBASZ II study, consumption of phytosterol-enriched foods was observed in a small proportion of the study group (less than 2% of participants). This translated into similar intakes of plant sterols and plant sterols along with phytosterols from fortified products.

Limitations

The main limitation of the study was its cross-sectional nature, as a result of which causal inferences cannot be drawn. Another limitation was the use of a single 24-h recall method, which may not reflect the usual pattern of food consumption. To reduce the possibility of bias, subjects who described their diet as atypical were excluded.

The strengths of this study are its representativeness to the Polish population and the assessment of diet quality, which may act synergistically with plant sterols, which has not been studied before. To minimize the synergistic effect of plant sterols on the association with CVD, the results were adjusted for the Healthy Diet Index (HDI) score. A strength of the study was its consideration of phytosterol-enriched foods.

5. Conclusions

This study suggests that habitual dietary intake of plant sterols may be associated with a lower chance of developing CVD, particularly in men. However, this finding should be treated with caution because of the difficulty in separating the effects of plant sterols from the effects of other dietary components that may have synergistic effects.

Author Contributions: Conceptualization, A.W. and A.M.W.; methodology, A.M.W., A.W. and M.E.Z.; software, A.W., A.M.W. and A.C.-M.; validation, A.W., A.M.W. and A.C.-M.; formal analysis, A.C.-M. and A.W.; investigation, A.W. and A.M.W.; resources, A.M.W., A.W., M.E.Z., I.M.-C. and W.D.; data curation, A.W. and A.M.W.; writing—original draft preparation, A.M.W.; writing—review and editing, A.W., M.E.Z., I.M.-C., A.C.-M. and W.D.; visualization, A.W., A.C.-M. and A.M.W.; supervision, A.W., A.M.W. and W.D.; project administration, A.M.W. and A.W.; funding acquisition, A.W. and A.M.W. All authors have read and agreed to the published version of the manuscript.

Funding: This research was funded by the National Institute of Cardiology (Grant no. 2.20/I/20) and Medical University of Bialystok (Grant no. SUB/1/DN/22/001/3317).

Institutional Review Board Statement: The study was conducted according to the guidelines of the Declaration of Helsinki and approved by the Bioethics Committee of the National Institute of Cardiology (protocol code 1344, date of approval 5 November 2012, and protocol code 1837, date of approval 14 January 2020).

Informed Consent Statement: Informed consent was obtained from all subjects involved in the study.

Data Availability Statement: All data in this study are made available upon request to the authors at the following e-mail address: anna.witkowska@umb.edu.pl or awaskiewicz@ikard.pl.

Conflicts of Interest: The authors declare no conflict of interest.

References

1. Roth, G.A.; Mensah, G.A.; Johnson, C.O.; Addolorato, G.; Ammirati, E.; Baddour, L.M.; Barengo, N.C.; Beaton, A.Z.; Benjamin, E.J.; GBD-NHLBI-JACC Global Burden of Cardiovascular Diseases Writing Group; et al. Global Burden of Cardiovascular Diseases and Risk Factors, 1990–2019: Update from the GBD 2019 Study. *J. Am. Coll. Cardiol.* **2020**, *76*, 2982–3021. [CrossRef] [PubMed]
2. Arnett, D.K.; Blumenthal, R.S.; Albert, M.A.; Buroker, A.B.; Goldberger, Z.D.; Hahn, E.J.; Himmelfarb, C.D.; Khera, A.; Lloyd-Jones, D.; McEvoy, J.W.; et al. 2019 ACC/AHA Guideline on the Primary Prevention of Cardiovascular Disease: A Report of the American College of Cardiology/American Heart Association Task Force on Clinical Practice Guidelines. *Circulation* **2019**, *140*, e596–e646. [CrossRef] [PubMed]
3. World Health Organization. *Prevention and Control of Noncommunicable Diseases: Guidelines for Primary Health Care in Low Resource Settings*; World Health Organization: Geneva, Switzerland, 2012.
4. Einarson, T.R.; Acs, A.; Ludwig, C.; Panton, U.H. Prevalence of cardiovascular disease in type 2 diabetes: A systematic literature review of scientific evidence from across the world in 2007–2017. *Cardiovasc. Diabetol.* **2018**, *17*, 83. [CrossRef] [PubMed]
5. Stamler, J.; Vaccaro, O.; Neaton, J.D.; Wentworth, D. Diabetes, other risk factors, and 12-yr cardiovascular mortality for men screened in the Multiple Risk Factor Intervention Trial. *Diabetes Care* **1993**, *16*, 434–444. [CrossRef]
6. Newman, J.D.; Schwartzbard, A.Z.; Weintraub, H.S.; Goldberg, I.J.; Berger, J.S. Primary Prevention of Cardiovascular Disease in Diabetes Mellitus. *J. Am. Coll. Cardiol.* **2017**, *70*, 883–893. [CrossRef]
7. Lloyd-Jones, D.M.; Morris, P.B.; Ballantyne, C.M.; Birtcher, K.K.; Daly, D.D., Jr.; DePalma, S.M.; Minissian, M.B.; Orringer, C.E.; Smith, S.C., Jr. 2017 Focused Update of the 2016 ACC Expert Consensus Decision Pathway on the Role of Non-Statin Therapies for LDL-Cholesterol Lowering in the Management of Atherosclerotic Cardiovascular Disease Risk: A Report of the American College of Cardiology Task Force on Expert Consensus Decision Pathways. *J. Am. Coll. Cardiol.* **2017**, *70*, 1785–1822. [CrossRef]
8. Gylling, H.; Simonen, P. Phytosterols, Phytostanols, and Lipoprotein Metabolism. *Nutrients* **2015**, *7*, 7965–7977. [CrossRef]
9. Ras, R.T.; Geleijnse, J.M.; Trautwein, E.A. LDL cholesterol-lowering effect of plant sterols and stanols across different dose ranges: A meta-analysis of randomised controlled studies. *Br. J. Nutr.* **2014**, *112*, 214–219. [CrossRef]
10. Klingberg, S.; Ellegård, L.; Johansson, I.; Jansson, J.H.; Hallmans, G.; Winkvist, A. Dietary intake of naturally occurring plant sterols is related to a lower risk of a first myocardial infarction in men but not in women in northern Sweden. *J. Nutr.* **2013**, *143*, 1630–1635. [CrossRef]
11. Pereira, T.S.; Fonseca, F.A.H.; Fonseca, M.I.H.; Martins, C.M.; Fonseca, H.A.R.; Fonzar, W.T.; Goulart, A.C.; Bensenor, I.M.; Lotufo, P.A.; Izar, M.C. Phytosterol consumption and markers of subclinical atherosclerosis: Cross-sectional results from ELSA-Brasil. *Nutr. Metab. Cardiovasc. Dis.* **2021**, *31*, 1756–1766. [CrossRef]
12. Witkowska, A.M.; Waśkiewicz, A.; Zujko, M.E.; Mirończuk-Chodakowska, I.; Cicha-Mikołajczyk, A.; Drygas, W. Assessment of Plant Sterols in the Diet of Adult Polish Population with the Use of a Newly Developed Database. *Nutrients* **2021**, *13*, 2722. [CrossRef] [PubMed]
13. Sirirat, R.; Heskey, C.; Haddad, E.; Tantamango-Bartley, Y.; Fraser, G.; Mashchak, A.; Jaceldo-Siegl, K. Comparison of phytosterol intake from FFQ with repeated 24-h dietary recalls of the Adventist Health Study-2 calibration sub-study. *Br. J. Nutr.* **2019**, *121*, 1424–1430. [CrossRef] [PubMed]
14. Ras, R.T.; van der Schouw, Y.T.; Trautwein, E.A.; Sioen, I.; Dalmeijer, G.W.; Zock, P.L.; Beulens, J.W. Intake of phytosterols from natural sources and risk of cardiovascular disease in the European Prospective Investigation into Cancer and Nutrition-the Netherlands (EPIC-NL) population. *Eur. J. Prev. Cardiol.* **2015**, *22*, 1067–1075. [CrossRef] [PubMed]
15. Yang, R.; Xue, L.; Zhang, L.; Wang, X.; Qi, X.; Jiang, J.; Yu, L.; Wang, X.; Zhang, W.; Zhang, Q.; et al. Phytosterol Contents of Edible Oils and Their Contributions to Estimated Phytosterol Intake in the Chinese Diet. *Foods* **2019**, *8*, 334. [CrossRef]
16. Wang, M.; Huang, W.; Hu, Y.; Zhang, L.; Shao, Y.; Wang, M.; Zhang, F.; Zhao, Z.; Mei, X.; Li, T.; et al. Phytosterol Profiles of Common Foods and Estimated Natural Intake of Different Structures and Forms in China. *J. Agric. Food Chem.* **2018**, *66*, 2669–2676. [CrossRef]
17. USDA. Composition of Foods, Raw, Processed, Prepared. National Nutrient Database for Standard Reference Release 28. Modified in 2019. 2015. Available online: https://data.nal.usda.gov/dataset/composition-foods-raw-processed-prepared-usda-national-nutrient-database-standard-reference-release-28-0 (accessed on 28 April 2022).
18. European Food Safety Authority (EFSA). Consumption of Food and Beverages with Added Plant Sterols. *EFSA J.* **2008**, *6*, 133.
19. Piepoli, M.F.; Hoes, A.W.; Agewall, S.; Albus, C.; Brotons, C.; Catapano, A.L.; Cooney, M.T.; Corrà, U.; Cosyns, B.; ESC Scientific Document Group; et al. 2016 European Guidelines on cardiovascular disease prevention in clinical practice: The Sixth Joint Task Force of the European Society of Cardiology and Other Societies on Cardiovascular Disease Prevention in Clinical Practice (constituted by representatives of 10 societies and by invited experts) developed with the special contribution of the European Association for Cardiovascular Prevention & Rehabilitation (EACPR). *Eur. Heart J.* **2016**, *37*, 2315–2381. [CrossRef]
20. Turini, E.; Sarsale, M.; Petri, D.; Totaro, M.; Lucenteforte, E.; Tavoschi, L.; Baggiani, A. Efficacy of Plant Sterol-Enriched Food for Primary Prevention and Treatment of Hypercholesterolemia: A Systematic Literature Review. *Foods* **2022**, *11*, 839. [CrossRef]

21. Catapano, A.L.; Graham, I.; De Backer, G.; Wiklund, O.; Chapman, M.J.; Drexel, H.; Hoes, A.W.; Jennings, C.S.; Landmesser, U.; Pedersen, T.R.; et al. 2016 ESC/EAS Guidelines for the management of dyslipidaemias. *Eur. Heart J.* **2016**, *37*, 2999–3058. [CrossRef]
22. Drygas, W.; Niklas, A.A.; Piwońska, A.; Piotrowski, W.; Flotyńska, A.; Kwaśniewska, M.; Nadrowski, P.; Puch-Walczak, A.; Szafraniec, K.; Bielecki, W.; et al. Multi-center National Population Health Examination Survey (WOBASZ II study): Assumptions, methods and implementation. *Kardiol. Pol.* **2016**, *74*, 681–690. [CrossRef]
23. Mendis, S.; Puska, P.; Norrving, B. *Global Atlas on Cardiovascular Disease Prevention and Control*; World Health Organization in Collaboration with the World Heart Federation and the World Stroke Organization: Geneva, Switzerland, 2011; pp. 3–18.
24. American Diabetes Association. 2. Classification and diagnosis of diabetes: Standards of Medical Care in Diabetes-2019. *Diabetes Care* **2019**, *42*, S13–S28. [CrossRef] [PubMed]
25. Kunachowicz, H.; Nadolna, I.; Przygoda, B.; Iwanow, K. *Food Composition Tables*; PZWL Medical Publishing House: Warsaw, Poland, 2005.
26. Zujko, M.E.; Witkowska, A.M. Antioxidant potential and polyphenol content of selected food. *Int. J. Food Prop.* **2011**, *14*, 300–308. [CrossRef]
27. Zujko, M.E.; Witkowska, A.M. Antioxidant potential and polyphenol content of beverages, chocolates, nuts and seeds. *Int. J. Food Prop.* **2014**, *17*, 86–92. [CrossRef]
28. Carlsen, M.H.; Halvorsen, B.L.; Holte, K.; Bøhn, S.K.; Dragland, S.; Sampson, L.; Willey, C.; Senoo, H.; Umezono, Y.; Sanada, C.; et al. The total antioxidant content of more than 3100 foods, beverages, spices, herbs and supplements used worldwide. *Nutr. J.* **2010**, *9*, 3. [CrossRef] [PubMed]
29. Neveu, V.; Perez-Jiménez, J.; Vos, F.; Crespy, V.; du Chaffaut, L.; Mennen, L.; Knox, C.; Eisner, R.; Cruz, J.; Wishart, D.; et al. Phenol-Explorer: An online comprehensive database on polyphenol contents in foods. *Database* **2010**, *2010*, bap024. [CrossRef] [PubMed]
30. Fransen, H.P.; Beulens, J.W.; May, A.M.; Struijk, E.A.; Boer, J.M.; de Wit, G.A.; Onland-Moret, N.C.; van der Schouw, Y.T.; Bueno-de-Mesquita, H.B.; Hoekstra, J.; et al. Dietary patterns in relation to quality-adjusted life years in the EPIC-NL kohort. *Prev. Med.* **2015**, *77*, 119–124. [CrossRef] [PubMed]
31. World Health Organization. *Diet, Nutrition, and the Prevention of Chronic Diseases*; WHO Technical Report Series No. 797, Report of a WHO Study Group; World Health Organization: Geneva, Switzerland, 1990.
32. Jóźwiak, J.J.; Studziński, K.; Tomasik, T.; Windak, A.; Mastej, M.; Catapano, A.L.; Ray, K.K.; Mikhailidis, D.P.; Toth, P.P.; LIPIDOGRAM2015 Investigators; et al. The prevalence of cardiovascular risk factors and cardiovascular disease among primary care patients in Poland: Results from the LIPIDOGRAM2015 study. *Atheroscler. Suppl.* **2020**, *42*, e15–e24. [CrossRef]
33. Jaceldo-Siegl, K.; Lütjohann, D.; Sirirat, R.; Mashchak, A.; Fraser, G.E.; Haddad, E. Variations in dietary intake and plasma concentrations of plant sterols across plant-based diets among North American adults. *Mol. Nutr. Food Res.* **2017**, *61*, 1600828. [CrossRef]
34. Wang, J.F.; Zhang, H.M.; Li, Y.Y.; Xia, S.; Wei, Y.; Yang, L.; Wang, D.; Ye, J.J.; Li, H.X.; Yuan, J.; et al. A combination of omega-3 and plant sterols regulate glucose and lipid metabolism in individuals with impaired glucose regulation: A randomized and controlled clinical trial. *Lipids Health Dis.* **2019**, *18*, 106. [CrossRef]
35. Liput-Sikora, A.; Cybulska, A.M.; Fabian, W.; Fabian-Danielewska, A.; Stanisławska, M.; Kamińska, M.S.; Grochans, E. Cardiovascular Risk Distribution in a Contemporary Polish Collective. *Int. J. Environ. Res. Publ. Health* **2020**, *17*, 3306. [CrossRef]
36. Glenn, A.J.; Lo, K.; Jenkins, D.J.A.; Boucher, B.A.; Hanley, A.J.; Kendall, C.W.C.; Manson, J.E.; Vitolins, M.Z.; Snetselaar, L.G.; Liu, S.; et al. Relationship Between a Plant-Based Dietary Portfolio and Risk of Cardiovascular Disease: Findings from the Women's Health Initiative Prospective Cohort Study. *J. Am. Heart Assoc.* **2021**, *10*, e021515. [CrossRef] [PubMed]
37. Jenkins, D.J.A.; Kendall, C.W.C.; Marchie, A.; Faulkner, D.; Vidgen, E.; Lapsley, K.G.; Trautwein, E.A.; Parker, T.L.; Josse, R.G.; Leiter, L.A.; et al. The effect of combining plant sterols, soy protein, viscous fibers, and almonds in treating hypercholesterolemia. *Metabolism* **2003**, *52*, 1478–1483. [CrossRef]
38. Jenkins, D.J.; Chiavaroli, L.; Wong, J.M.; Kendall, C.; Lewis, G.F.; Vidgen, E.; Connelly, P.W.; Leiter, L.A.; Josse, R.G.; Lamarche, B. Adding monounsaturated fatty acids to a dietary portfolio of cholesterol-lowering foods in hypercholesterolemia. *CMAJ* **2010**, *182*, 1961–1967. [CrossRef]
39. Chiavaroli, L.; Nishi, S.K.; Khan, T.A.; Braunstein, C.R.; Glenn, A.J.; Mejia, S.B.; Rahelić, D.; Kahleová, H.; Salas-Salvadó, J.; Jenkins, D.J.A.; et al. Portfolio Dietary Pattern and Cardiovascular Disease: A Systematic Review and Meta-analysis of Controlled Trials. *Prog. Cardiovasc. Dis.* **2018**, *61*, 43–53. [CrossRef]
40. Andersson, S.W.; Skinner, J.; Ellegård, L.; Welch, A.A.; Bingham, S.; Mulligan, A.; Andersson, H.; Khaw, K.T. Intake of dietary plant sterols is inversely related to serum cholesterol concentration in men and women in the EPIC Norfolk population: A cross-sectional study. *Eur. J. Clin. Nutr.* **2004**, *58*, 1378–1385. [CrossRef] [PubMed]
41. Wang, P.; Chen, Y.M.; He, L.P.; Chen, C.G.; Zhang, B.; Xue, W.Q.; Su, Y.X. Association of natural intake of dietary plant sterols with carotid intima-media thickness and blood lipids in Chinese adults: A cross-section study. *PLoS ONE* **2012**, *7*, e32. [CrossRef] [PubMed]
42. Klingberg, S.; Ellegard, L.; Johansson, I.; Hallmans, G.; Weinehall, L.; Andersson, H.; Winkvist, A. Inverse relation between dietary intake of naturally occurring plant sterols and serum cholesterol in northern Sweden. *Am. J. Clin. Nutr.* **2008**, *87*, 993–1001. [CrossRef]

Article

Association between Diet Quality and Index of Non-Alcoholic Steatohepatitis in a Large Population of People with Type 2 Diabetes: Data from the TOSCA.IT Study

Marilena Vitale [1,†], Giuseppe Della Pepa [1,†], Giuseppina Costabile [1], Lutgarda Bozzetto [1], Paola Cipriano [1], Stefano Signorini [2], Valerio Leoni [2,3], Gabriele Riccardi [1], Olga Vaccaro [1,*] and Maria Masulli [1]

1 Department of Clinical Medicine and Surgery, University of Naples Federico II, 80131 Naples, Italy
2 Laboratory of Clinical Biochemistry, Hospital Pius XI of Desio, ASST-Brianza, 20833 Desio, Italy
3 Department of Medicine and Surgery, University of Milano-Bicocca, 20216 Monza, Italy
* Correspondence: ovaccaro@unina.it; Tel.: +39-081-746-3665
† These authors contributed equally to this work.

Abstract: Background: There are still open questions with respect to the optimal dietary treatment in patients with type 2 diabetes (T2D) and coexisting non-alcoholic steatohepatitis (NASH). The aim of this study is to investigate, in patients with T2D, the association between NASH, dietary component intake, food groups and adherence to the Mediterranean diet. Methods: Cross-sectional analysis of 2026 people with T2D (1136 men and 890 women). The dietary habits were assessed with the European Prospective Investigation into Cancer and Nutrition (EPIC) questionnaire. NASH was identified by the Index Of NASH (ION). Based on the cluster analysis two dietary patterns were identified: the NASH and the NO-NASH pattern. Results: The macronutrient composition of the diet was similar in the two patterns. However, the NASH pattern compared with the NO-NASH pattern was characterized by a significantly lower content of fibre ($p < 0.001$), β-carotene ($p < 0.001$), vitamin C ($p < 0.001$), vitamin E ($p < 0.001$), polyphenols ($p = 0.026$) and antioxidant capacity ($p < 0.001$). With regard to food consumption, the NASH pattern compared with NO-NASH pattern was characterized by higher intake of rice ($p = 0.021$), potatoes ($p = 0.013$), red ($p = 0.004$) and processed meat ($p = 0.003$), and a lower intake of wholegrain bread ($p = 0.019$), legumes and nuts ($p = 0.049$), vegetables ($p = 0.047$), fruits ($p = 0.002$), white meat ($p = 0.001$), fatty fish ($p = 0.005$), milk and yogurt ($p < 0.001$). Conclusions: NO-NASH dietary pattern was characterized by a food consumption close to the Mediterranean dietary model, resulting in a higher content of polyphenols, vitamins, and fibre. These finding highlight the potential for dietary components in the prevention/treatment of NASH in people with T2D.

Keywords: type 2 diabetes; NASH; micronutrients; macronutrients; dietary habits; foods groups; Mediterranean diet; dietary patterns

Citation: Vitale, M.; Della Pepa, G.; Costabile, G.; Bozzetto, L.; Cipriano, P.; Signorini, S.; Leoni, V.; Riccardi, G.; Vaccaro, O.; Masulli, M. Association between Diet Quality and Index of Non-Alcoholic Steatohepatitis in a Large Population of People with Type 2 Diabetes: Data from the TOSCA.IT Study. *Nutrients* **2022**, *14*, 5339. https://doi.org/10.3390/nu14245339

Academic Editor: Henry J. Thompson

Received: 22 November 2022
Accepted: 12 December 2022
Published: 15 December 2022

Publisher's Note: MDPI stays neutral with regard to jurisdictional claims in published maps and institutional affiliations.

Copyright: © 2022 by the authors. Licensee MDPI, Basel, Switzerland. This article is an open access article distributed under the terms and conditions of the Creative Commons Attribution (CC BY) license (https://creativecommons.org/licenses/by/4.0/).

1. Introduction

Non-alcoholic fatty liver disease (NAFLD) represents the most common liver disease worldwide affecting approximatively 20–30% of the general population [1]. The histopathological and clinical abnormalities of NAFLD spectrum ranges from the accumulation of triglycerides in the liver, i.e., non-alcoholic fatty liver (NAFL), to the inflammation and cellular damage of the hepatocytes, i.e., non-alcoholic steatohepatitis (NASH), that may progress to liver fibrosis and advanced cirrhosis [2]. The most serious clinical manifestations of NAFLD, i.e., NASH and cirrhosis, have very recently become the fastest growing indications for liver transplantation in western countries, heavily impacting on patient health, economic aspects and quality of life [3].

Interestingly, NAFLD is strictly associated with the features of metabolic syndrome such as obesity, type 2 diabetes (T2D), dyslipidaemia and hypertension [4]. In particular, the association between NAFL/NASH and T2D is well established and there appears to exist an intricate interrelationship whereby the existence of one drives progression to the other. T2D seems to be the most important risk factor for NAFLD and the most important clinical predictor of the advanced forms of NAFLD [5,6]. On the other hand, NAFLD is associated with a worse metabolic profile [7,8] and a higher prevalence of microvascular and macrovascular complications of diabetes, independently of other known risk factors [9–11]. From an epidemiological point of view, it is not surprising that there is a high prevalence of NAFL and NASH in T2D, estimated at 55–70% and 20–40%, respectively [12], and higher in T2D with obesity [13].

Although liver biopsy represents the gold standard for the diagnosis of NASH, it is not feasible in large epidemiological studies. Several indices, based on non-invasive measures easily performed in clinical practice, have been proposed for the diagnosis of NASH [14], although none of these has been validated in people with diabetes. Among others, we used the Index Of NASH (ION), an algorithm constructed from the combination of triglycerides, visceral obesity, alanine aminotransferase (ALT) and Homeostatic Model Assessment (HOMA-IR), and validated against liver biopsy in an obese population sharing several metabolic and clinical features with T2D (i.e., obesity, excess of visceral fat, insulin resistance and high prevalence of NASH) [15]; in this population, the ION has shown a good diagnostic accuracy (AUC = 0.88 [95%CI 0.82–0.95]), with a sensitivity of 92% and a specificity of 60% [15].

No pharmaceutical approaches for NAFLD have been approved to date, and the cornerstone in the prevention and treatment of NAFLD and its severe forms is represented by lifestyle modifications, including diet-related factors [14,16]. Some attempts have been made to clarify the association between dietary components and NAFLD in the general population. Outside the context of clinical trials, epidemiological studies show that high glycaemic index foods and intake in saturated fats and simple sugars—fructose in particular—are associated with a higher prevalence in NAFLD [17–21]; whereas the intake of *n*-3 and *n*-6 polyunsaturated fatty acids (PUFA), monounsaturated fatty acids (MUFA), and fibre [22] appears to be associated with a lower prevalence of NAFLD [18,21,22]. Among micronutrients, intake of vitamins [23] and polyphenols is associated with a lower prevalence of NAFLD and might beneficially impact on the progression from NAFL to NASH [24].

In individuals with coexisting T2D and NAFL/NASH, hypocaloric diets promoting a weight loss of 7–10% are effective in improving metabolic parameters of both diseases [14], but they are not feasible in the long term and the optimal dietary model for people with T2D and NASH, not subjected to caloric restriction, remains ill-defined [25,26]. Nutritional guidelines for the treatment of diabetes recommend 45–60% of carbohydrates, selecting those with a low glycaemic index and high in fibre, 25–35% of fats, preferring MUFA and PUFA, 15–20% of proteins, and limiting/avoiding the intake of free sugars, sugar-sweetened beverages and added fructose [26,27]. These recommendations are designed with the main focus on correction of hyper-glycaemia; furthermore, the patient's compliance is generally low/very low. Indications based on food consumption, rather than on nutrients, may improve adherence, but evidence regarding the possible association between habitual food consumption and NASH in T2D is lacking.

The aim of this study is to investigate in a large population of patients with T2D the association of habitual diet (i.e., diet composition and food consumption) with NASH, in order to expand knowledge on the potential for dietary components in the prevention/treatment of NASH in people with T2D.

2. Materials and Methods

2.1. Study Design and Population

We conducted a cross sectional study within the framework of the TOSCA.IT study (NCT00700856), a randomized controlled trial designed to evaluate the effects of sulfonylurea or pioglitazone, in add-on to metformin, on cardiovascular events in people with T2D.

For the aim of the present study, we used data collected at baseline, prior to randomization to the study treatments.

The study participants were people with T2D, aged 50–75 years, on stable treatment with a full dose of metformin (2–3 g per day), and with a glycated hemoglobin (HbA1c) between 7% and 9%. Participants with incomplete data sets, those with alcohol intake exceeding 30 g/day if men and 20 g/day if women, or taking *n*-3 supplements were excluded from the analyses [28]; other exclusion criteria were severe hepatic dysfunction (plasma ALT values > 2.5 times the upper normal limit), serum creatinine > 1.5 mg/dL, history of congestive heart failure, (NYHA class I or higher), ulcer or gangrene of the lower extremities, cancer, substance abuse, and any health problem requiring special dietary treatments. Details of the study protocol have been published [29,30]. NASH was defined based on ION $\geq$ 50. To identify the association of the habitual diet with NASH, dietary patterns associated with ION $\geq$ 50 or <50 were derived using the K-means cluster analysis by which the sample population is classified into homogenous groups presenting different characteristics using a specific variable as the comparison criterion, in our case the ION. To perform this analysis, firstly the 248 food items were categorized into 59 food groups based on their similarity in term of nutrient composition. Then, the K-means clustering method was performed and the algorithm utilized to identify within each cluster the smallest variation. Two clusters were produced using a non-hierarchical K-means clustering method, with the random seed and 10 iterations to refine and optimize the classifications, and participants were grouped according to Euclidean distances. Two clusters were identified, one associated with an ION $\geq$ 50, the other associated with an ION < 50, respectively defined in the text and tables as cluster NASH or cluster NO-NASH. The anthropometric, metabolic and nutritional variables were compared in these two patterns. The study protocol has been approved by the Ethics Review Committee of the Coordinating Center and of each participating center. All participant provided written informed consent before entering the study.

2.2. Assessment of Anthropometric and Laboratory Parameters

Body weight was measured by mechanic balance (Seca 721), height with bar-altimeter, waist and hip circumference using an anelastic meter. Waist circumference was measured halfway between the lower ribs and the iliac crest and hip circumference was measured at the largest point around the buttocks. All measures were taken with an accuracy to the nearest 0.1 kg and 0.1 cm, respectively, and with the patient in light clothing and no shoes. Body mass index (BMI) was calculated as weight (kg)/height (m^2). Blood pressure was measured in the sitting position by a standard protocol. Blood samples were obtained in the morning after an overnight fast and all biochemical parameters were analyzed in a central laboratory. Plasma glucose, total cholesterol, HDL-cholesterol, triglycerides, liver enzymes—ALT, aspartate aminotransferase (AST), and gamma-glutamyl-transpeptidase (GGT)—and high sensitivity C-reactive protein were detected by standard methods. LDL-cholesterol was calculated according to the Friedewald equation for triglyceride values < 400 mg/dL. HbA1c was measured with high liquid performance chromatography standardized according to IFCC. Plasma insulin was detected by ELISA (DIAsource ImmunoAssays S.A., Nivelles, Belgium) on a Triturus Analyser (Diagnostics Grifols, S.A., Barcelona, Spain). Insulin resistance was evaluated by the HOMA method calculated as follows: fasting glucose (mg/dL) $\times$ fasting insulin (μU/mL)/405.

2.3. Assessment of Dietary Intake, Food Consumption and Adherence to the Mediterranean Diet

The evaluation of eating habits was performed through the Italian version of the European Prospective Investigation into Cancer and Nutrition (EPIC) questionnaire, a validated food frequency questionnaire (FFQ) for the assessment of dietary habits in large epidemiological studies [31,32]. Details have been reported elsewhere [33,34]. Briefly, the FFQ is based on 248 items for which the respondent has to report (1) the absolute frequency of consumption in terms of per day, week, month, or year, and (2) the quantity by selection of pictures showing the portion size as small, medium and large, with additional quantifiers (i.e., "smaller than the small portion" or "between the small and medium portion", etc ...). Incomplete questionnaires and questionnaires with energy intake less than 800 or greater than 5000 kcal/day were excluded. A specific software (Nutrition Analysis of food frequency questionnaire—FFQ) was used to obtain the average daily amounts of foods (g/day) [31,32] and the energy and nutrient composition of the habitual diet [35,36]. The intake of polyphenols was evaluated using the USDA database [37] in combination with the Phenol-Explorer®database [38], as reported in more detail elsewhere [33,34]. In order to evaluate the adherence to the Mediterranean Diet, the relative Mediterranean Diet (rMED) score was used [39] as described in a prior publication [40]. Briefly, the average daily intake of fruits, vegetables, cereals, legumes, fish, olive oil, meat and meat products, dairy products and alcohol was divided in tertiles and assigned a score of 0, 1 or 2 to the first, second or third tertile, respectively, for the groups fitting the Mediterranean model, whereas for meat and dairy products, we assigned a score of 0, 1 or 2 to the third, second and first tertile, respectively. Regarding alcohol intake, 2 points for moderate intake (i.e., 5–25 g/day for women and 10–50 g/day for men, respectively) and 0 points for a consumption at or below the sex-specific range were assigned. The rMED score ranged from 0 to 18.

2.4. Assessment of Non-Alcoholic Steatohepatitis

NASH was calculated as indirect index by the ION according to the following formula: 1.33 waist to hip ratio + 0.03 × triacyclglycerols (mg/dL) + 0.18 × ALT (U/l) + 8.53 × HOMA − 13.93 for men; 0.02 × triacyclglycerols (mg/dL) + 0.24 × ALT (U/l) + 9.61 × HOMA − 13.99 for women. NASH was identified by an ION score of ≥50 [15].

2.5. Statistical Analysis

Data are presented as means ± standard deviation, frequencies and percentages, as appropriate. The *t*-test for independent samples was used to compare group means; for skewed variables, log transformed values were used. The χ^2 test vas used to compare frequencies. A *p*-value < 0.05, two-tailed, was considered significant and all analyses were conducted with the SPSS Statistics software 28.0 (SPSS/PC; IBM, Armonk, NY, USA).

3. Results

The study population consists of 2026 people with T2D (1136 men and 890 women) with a mean age of 62.1 ± 6.5 years, a mean BMI of 30.3 ± 4.5 kg/m^2 and a mean duration of diabetes of 8.5 ± 5.7 years. The prevalence of NASH according to the ION was 32%.

In Table 1 are reported the general characteristics and the cardio-metabolic profile for participants in the two clusters. The BMI, waist and hip circumference, waist/hip ratio, systolic and diastolic blood pressure, HbA1c, fasting plasma glucose and insulin, HOMA-IR, plasma LDL-cholesterol, plasma triglycerides, and liver enzymes were significantly higher among people in the cluster NASH as compared with those in the cluster NO-NASH, while age, diabetes duration and plasma HDL-cholesterol values were significantly lower. The proportion of smokers was similar in the two clusters, and a high proportion of the population was taking lipid lowering medications (62%) and/or antihypertensive medications (93%), with no significant differences between the two clusters.

In Table 2 are reported the energy intake and the nutrient composition of the diet in the participants in the two clusters. A significantly lower intake of energy, fibre, vitamin C, ß-carotene, vitamin E and polyphenols, was observed in the cluster NASH; accordingly the antioxidant capacity of the diet, estimated as Trolox Equivalent Antioxidant Capacity (TEAC), Total Radical-Trapping Antioxidant Parameter (TRAP), Ferric Reducing-Antioxidant Power (FRAP) was significantly lower. No differences were detected for the other components of the diet between the groups.

Table 1. General characteristics and metabolic profile in the clusters NASH or NO-NASH.

	Cluster NASH (n. 642)	Cluster NO-NASH (n. 1384)	*p*-Value
Age (years)	62 ± 7	63 ± 6	0.006
Smoking (%)	34.4	32.6	0.209
Diabetes duration (years)	8 ± 5	9 ± 6	0.004
BMI (kg/m^2)	32 ± 4	29 ± 4	<0.0001
Waist circumference (cm)	108 ± 11	102 ± 11	<0.0001
Hip circumference (cm)	109 ± 11	105 ± 10	<0.0001
Waist/Hip ratio	0.99 ± 0.98	0.96 ± 0.91	0.003
Systolic blood pressure (mm/Hg)	136 ± 16	134 ± 15	0.004
Diastolic blood pressure (mm/Hg)	81 ± 9	80 ± 9	0.002
HbA1c (%)	7.8 ± 0.5	7.6 ± 0.5	<0.0001
Plasma Glucose (mg/dL)	185 ± 39	159 ± 32	<0.0001
Plasma Insulin (µU/mL)	23.5 ± 12.2	9.4 ± 3.6	<0.0001
HOMA-IR	10.7 ± 7.0	3.6 ± 1.4	<0.0001
Plasma HDL-cholesterol (mg/dL)	43 ± 12	48 ± 12	<0.0001
Plasma LDL-cholesterol (mg/dL)	104 ± 30	101 ± 36	0.045
Plasma Triglycerides (mg/dL)	184 ± 114	137 ± 66	<0.0001
C-reactive protein (mg/dL)	0.5 ± 2.3	0.4 ± 1.7	0.246
eGFR (ml/min/1.73 m^2)	91.4 ± 2.6	92.7 ± 2.5	0.311
AST (U/L)	24 ± 12	18 ± 9	<0.0001
ALT (U/L)	25 ± 15	17 ± 10	<0.0001
GGT (U/L)	47 ± 54	31 ± 28	<0.0001
Use of antihypertensive drugs (%)	95.3	91.2	0.652
Use of Lipid lowering drugs (%)	63.6	61.1	0.441

Data are means ± SD. BMI: body mass index; HbA1c: glycated hemoglobin; HOMA-IR, homeostatic model assessment; eGFR: estimated glomerular filtration rate; AST: aspartate aminotransferase; ALT: alanine aminotransferase; GGT: gamma-glutamyl-transpeptidase.

Table 2. Energy intake and nutrient composition of the habitual diet in in the clusters NASH or NO-NASH.

	Cluster NASH (n. 642)	Cluster NO-NASH (n. 1384)	*p*-Value
Energy (Kcal/day)	1755 ± 618	1843 ± 692	0.006
Total Proteins (% TEI)	18.4 ± 2.6	18.2 ± 2.6	0.272
Proteins from animal food sources (% TEI)	12.7 ± 3.2	12.5 ± 3.2	0.276
Proteins from vegetable food sources (% TEI)	5.7 ± 1.1	5.7 ± 1.1	0.574
Total Lipids (% TE)	37.1 ± 6.2	36.6 ± 6.1	0.089
Saturated fatty acids (% TEI)	12.2 ± 2.5	12.0 ± 2.5	0.179
Monounsaturated fatty acids (% TEI)	18.3 ± 3.8	18.0 ± 3.8	0.105
Polyunsaturated fatty acids (% TEI)	4.5 ± 1.1	4.5 ± 1.1	0.253
n-3 (% TEI)	0.55 ± 0.11	0.55 ± 0.12	0.930
n 6 (% TEI)	3.60 ± 1.03	3.56 ± 1.03	0.394
Total cholesterol (mg/1000 kcal/day)	185 ± 53	181 ± 54	0.130
Total Carbohydrates (% TEI)	44.4 ± 7.7	45.1 ± 7.4	0.081
Added sugars (% TEI)	2.37 ± 3.01	2.24 ± 3.24	0.376
Fibre (g/1000 kcal/day)	10.5 ± 2.6	11.0 ± 2.7	<0.0001
Glycemic Index	51.6 ± 3.5	51.9 ± 3.4	0.172
Glycemic Load (%)	105.1 ± 46.6	111.4 ± 50.1	0.019
Alcohol (g/day)	9.9 ± 15.2	11.0 ± 15.3	0.146
Vitamin-C (mg/day)	105 ± 49	115 ± 58	<0.0001
β-carotene (mg/day)	2286 ± 1307	2649 ± 1830	<0.0001
Vitamin E (mg/day)	6.42 ± 2.27	7.00 ± 2.90	<0.0001
Vitamin D (mg/day)	2.47 ± 1.29	2.53 ± 1.53	0.398
TEAC	5.47 ± 2.25	6.00 ± 2.42	<0.0001
TRAP	8.15 ± 3.64	8.91 ± 3.76	<0.0001
FRAP	17.04 ± 7.11	18.46 ± 7.45	<0.0001
Total polyphenols (mg/1000 kcal/day)	377.4 ± 163.1	386.1 ± 165.4	0.026

Data are means ± SD. TEI: total energy intake; TEAC: trolox equivalent antioxidant capacity; TRAP: total radical-trapping antioxidant parameter; FRAP: ferric reducing-antioxidant power.

Coherent with these findings, the rMED score, an indicator of the overall quality of the adherence to the Mediterranean dietary pattern, was significantly lower in the cluster NASH (Table 3). This data was due to significant differences in individual foods and food group consumption (Table 3). People in the cluster NASH showed a significantly higher consumption of pasta, rice, potatoes, total meat, red meat, processed meat and a lower consumption of wholegrain bread, legumes and nuts, vegetables, fruits, white meat, fatty fish, total dairy products, milk and yogurt. The consumption of total cereals, white bread, total fish, lean fish, cheese, eggs, vegetable oils and fats from animal origin was not significantly different between the groups.

Table 3. Consumption of specific food items (g/1000 Kcal/day), in the clusters NASH or NO-NASH.

	Cluster NASH (n. 642)	Cluster NO-NASH (n. 1384)	*p*-Value
Cereals	97.9 ± 35.1	94.8 ± 36.6	0.075
Pasta	28.2 ± 17.7	26.8 ± 16.3	0.016
Rice	3.55 ± 4.36	3.09 ± 4.04	0.021
Bread	35.6 ± 31.1	36.1 ± 32.6	0.738
Wholegrain Bread	6.7 ± 15.8	8.7 ± 17.6	0.019
Legumes and Nuts	13.6 ± 10.6	14.7 ± 11.8	0.049
Vegetables	95.7 ± 49.6	97.4 ± 49.7	0.047
Potatoes	11.7 ± 12.5	10.1 ± 14.0	0.013
Fruits	151.1 ± 86.1	164.3 ± 92.9	0.002
Meat	73.7 ± 30.7	68.4 ± 29.1	<0.0001
Red meat	56.3 ± 26.8	52.7 ± 25.8	0.004
White meat	1.5 ± 0.9	1.9 ± 0.5	0.001
Processed meat	15.6 ± 11.2	14.0 ± 10.1	0.003
Fish	23.7 ± 18.8	23.0 ± 17.1	0.390
Fatty fish	14.6 ± 12.1	16.3 ± 11.8	0.005
Lean fish	8.5 ± 6.0	8.2 ± 6.5	0.147
Dairy products	100.4 ± 54.9	125.3 ± 60.1	<0.0001
Cheese	20.1 ± 13.0	20.0 ± 13.4	0.965
Milk and Yogurt	80.6 ± 76.0	102.9 ± 88.0	<0.0001
Eggs	10.6 ± 7.2	10.4 ± 7.5	0.543
Vegetable oils	14.0 ± 6.0	13.9 ± 6.0	0.897
Oils from animal origin	1.56 ± 1.54	1.43 ± 1.42	0.067
rMED score	8.1 ± 2.6	9.7 ± 2.5	0.001

Data are means ± SD. rMED: relative Mediterranean Diet.

4. Discussion

To the best of our knowledge, this is the first large cross-sectional study evaluating the association between NASH and habitual diet in a well-characterized sample of adults with T2D.

The prevalence of NASH in the study population was high (32%) and in line with recent epidemiological data, as was the more adverse cardiovascular risk factors profile observed in the cluster NASH [12]. The association between nutrient composition of the diet and NAFLD, particularly NAFL, has been reported in prior epidemiological studies [41], but less is known regarding the association with more advanced stages of NAFLD (i.e., NASH) and no data are available on the association between habitual diet and NASH in patients with T2D. Here, we provide data on this association, thus expanding current knowledge.

The first remarkable finding is that the nutrient distribution was largely similar in the NASH or NO-NASH cluster. This is at variance with studies assessing the relation between NAFDL and diet composition in non-diabetic people which describe higher intakes of cholesterol, saturated fat [42,43] and added sugars [44,45] in individuals with NAFLD in comparison with matched controls without NAFLD. This inconsistency might be related to differences in the study design; in fact, we studied a population consisting of patients with T2D regularly attending diabetes clinics and who, therefore, may have restricted the consumption of the above-mentioned items as a result of medical advice, thus diluting the association.

The second relevant finding of this study is the association between fibre, micronutrient intake and NASH. Interestingly, despite the composition in macronutrients being similar between groups, the overall quality of the diet was very different between groups; with this regard, the intake of fibre, vitamins and polyphenols was significantly lower in the cluster NASH. These results are in line with epidemiological and clinical studies performed in patients without T2D [23,46,47]. All together, available evidence suggests that vitamins and polyphenols might prevent the advance of steatosis to NASH, probably restoring oxidative stress and reducing the transcription of the pro-inflammatory cytokines, which are the main drivers in the progression from NAFL to NASH [24,48,49]. Furthermore, dietary fibre might positively influence NAFL by acting on postprandial metabolic state, decreasing glucose absorption with a consequent reduction of the hepatic influx of glucose and de-novo lipo-genesis [22]. In addition, dietary fibre might stimulate a healthy gut microbiota, consequently decreasing the development of tissue inflammation and liver injury that led to NASH [50,51] and expanding also to NASH the interest in bioactive food compounds for T2D.

In terms of food groups, the cluster NASH was characterized by a lower consumption of whole grain bread, legumes and nuts, vegetables, fruits, fatty fish, milk and yogurt and a higher consumption of pasta, rice, potatoes, red meat and processed meat, providing data on the relation between food groups, NASH and T2D in line with those available for people without diabetes [43,52,53].

These data indicate that overall the NO-NASH dietary pattern is close to the Mediterranean dietary model, contributing to the growing evidence suggesting this model as the reference nutritional pattern to prevent and treat NAFLD [14,54], also in people with T2D. The beneficial effects of the Mediterranean diet on NAFLD might be related to dietary components such as dietary fibre, polyphenols and vitamins, that lead to the enhancement in the most important risk factors of NAFLD, such as BMI, insulin resistance and serum triglycerides [54], which are also key pathogenic factors for the development of T2D and major determinants of blood glucose control once diabetes has developed.

To our knowledge, this is the first epidemiological study on a large population of patients with T2D to evaluate the association between NASH and different dietary components and food groups, in a real-life setting.

There are still many open questions with respect to dietary treatment in individuals with both NASH and T2D. Importantly, in spite of the intimate relation between NAFLD and T2D, there are few nutritional intervention trials in which patients with coexisting T2D and NASH have been included. Therefore, it is unclear whether results from the numerous trials performed in patients with NASH and without T2D can be generalized to patients with both diseases. This study, by including a detailed analysis of both vegetable-based and animal-based foods, raises hypotheses on the overall dietary approach for NASH in people with diabetes, which could require confirmation in future large randomized-controlled trials.

Some limitations of the study must be acknowledged. First, the causal relationship between dietary components, Mediterranean dietary score and NASH cannot be proven due to the cross-sectional study design. Second, potential confounding from unmeasured lifestyle factors, such as physical activity level, might exist. Furthermore, data regarding dietary habits were collected only once and, consequently, could be prone to seasonal fluctuation and recall bias. Finally, NASH was detected by an indirect index currently accepted by NAFLD guidelines [16,55]. This index, although not specifically developed for people with diabetes was, however, validated in people with obesity who share several metabolic and clinical features with T2D (obesity, excess of visceral fat, insulin resistance and high prevalence of NASH), [15].

These limitations are counterbalanced by several strengths: a large sample size, a well-defined population of patients with T2D studied within the context of real-life clinical practice, the collection of nutritional and clinical data according to standard methods and biochemical measurements performed in a centralized laboratory.

5. Conclusions

In conclusion, this is one of the first epidemiological studies to investigate the dietary correlates of NASH in free living people with T2D focusing on foods and food groups. The results provide insights regarding habitual food consumption and dietary components as correlates of NASH in people with coexisting diabetes, showing that the NO-NASH dietary pattern is characterized by a higher intake of whole grain-based foods, legumes, nuts, fruits, vegetables, fatty fish, milk and yogurt, translating into a higher intake of polyphenols, vitamins and fibre and a higher Mediterranean dietary score. These findings expand current knowledge by highlighting the potential for dietary components in the prevention/treatment of NASH in people with T2D.

Author Contributions: Conceptualization, M.V., G.D.P. and O.V.; methodology, M.V., G.C., P.C., L.B., S.S. and V.L.; formal analysis, M.V. and G.D.P.; data curation, M.V., G.D.P. and M.M.; writing—original draft preparation, M.V. and G.D.P.; writing—review and editing, M.M. and O.V.; supervision, O.V. and G.R.; funding acquisition, O.V. All authors have read and agreed to the published version of the manuscript.

Funding: The study was supported by the Italian Medicines Agency (AIFA) within the Independent Drug Research Program—contract number FARM6T9CET—and by Diabete Ricerca, the non-profit Research Foundation of the Italian Diabetes Society. The funding agency played no role in the study design; in the collection, analysis, and interpretation of data; in the writing of the manuscript; or in the decision to submit the manuscript for publication.

Institutional Review Board Statement: The study was conducted in accordance with the Declaration of Helsinki, and approved by the FEDERICO II UNIVERISTY Ethics Committee (protocol code: 123/08; date of approval: 21 January 2008).

Informed Consent Statement: Informed consent was obtained from all subjects involved in the study.

Data Availability Statement: Not applicable.

Acknowledgments: The participation of the patients in the study is gratefully acknowledged. We thank all the investigators and the dietitians in the TOSCA.IT centers for their excellent cooperation. We are also indebted to the administrative personnel of the Italian Diabetes Society (SID) for their support.

Conflicts of Interest: The authors declare no conflict of interest.

References

1. Cotter, T.G.; Rinella, M. Nonalcoholic Fatty Liver Disease 2020: The State of the Disease. *Gastroenterology* **2020**, *158*, 1851–1864. [CrossRef] [PubMed]
2. Lazarus, J.V.; Colombo, M.; Cortez-Pinto, H.; Huang, T.T.; Miller, V.; Ninburg, M.; Schattenberg, J.M.; Seim, L.; Wong, V.W.S.; Zelber-Sagi, S. NAFLD—Sounding the alarm on a silent epidemic. *Nat. Rev. Gastroenterol. Hepatol.* **2020**, *17*, 377–379. [CrossRef] [PubMed]
3. Younossi, Z.M.; Koenig, A.B.; Abdelatif, D.; Fazel, Y.; Henry, L.; Wymer, M. Global epidemiology of nonalcoholic fatty liver disease-Meta-analytic assessment of prevalence, incidence, and outcomes. *Hepatology* **2016**, *64*, 73–84. [CrossRef]
4. Targher, G. From nonalcoholic fatty liver disease to metabolic dysfunction-associated fatty liver disease: More than a single-letter change in an acronym. *Hepatoma Res.* **2021**, *7*, 47. [CrossRef]
5. Cusi, K. Time to Include Nonalcoholic Steatohepatitis in the Management of Patients with Type 2 Diabetes. *Diabetes Care* **2020**, *43*, 275–279. [CrossRef]
6. Gastaldelli, A.; Cusi, K. From NASH to diabetes and from diabetes to NASH: Mechanisms and treatment options. *JHEP Rep.* **2019**, *1*, 312–328. [CrossRef]
7. Lomonaco, R.; Bril, F.; Portillo-Sanchez, P.; Ortiz-Lopez, C.; Orsak, B.; Biernacki, D.; Lo, M.; Suman, A.; Weber, M.H.; Cusi, K. Metabolic Impact of Nonalcoholic Steatohepatitis in Obese Patients with Type 2 Diabetes. *Diabetes Care* **2016**, *39*, 632–638. [CrossRef]
8. Bril, F.; Cusi, K. Management of Nonalcoholic Fatty Liver Disease in Patients with Type 2 Diabetes: A Call to Action. *Diabetes Care* **2017**, *40*, 419–430. [CrossRef]
9. Targher, G.; Bertolini, L.; Rodella, S.; Tessari, R.; Zenari, L.; Lippi, G.; Arcaro, G. Nonalcoholic fatty liver disease is independently associated with an increased incidence of cardiovascular events in type 2 diabetic patients. *Diabetes Care* **2007**, *30*, 2119–2121. [CrossRef]
10. Targher, G.; Marra, F.; Marchesini, G. Increased risk of cardiovascular disease in non-alcoholic fatty liver disease: Causal effect or epiphenomenon? *Diabetologia* **2008**, *51*, 1947–1953. [CrossRef] [PubMed]
11. Hazlehurst, J.M.; Woods, C.; Marjot, T.; Cobbold, J.F.; Tomlinson, J.W. Non-alcoholic fatty liver disease and diabetes. *Metabolism* **2016**, *65*, 1096–1108. [CrossRef] [PubMed]

12. Stefan, N.; Cusi, K. A global view of the interplay between non-alcoholic fatty liver disease and diabetes. *Lancet Diabetes Endocrinol.* **2022**, *10*, 284–296. [CrossRef] [PubMed]
13. Diehl, A.M.; Day, C. Cause, Pathogenesis, and Treatment of Nonalcoholic Steatohepatitis. *N. Engl. J. Med.* **2017**, *377*, 2063–2072. [CrossRef] [PubMed]
14. European Association for the Study of the Liver (EASL); European Association for the Study of Diabetes (EASD); European Association for the Study of Obesity (EASO). EASL-EASD-EASO Clinical Practice Guidelines for the management of non-alcoholic fatty liver disease. *J. Hepatol.* **2016**, *64*, 1388–1402. [CrossRef]
15. Otgonsuren, M.; Estep, M.J.; Hossain, N.; Younossi, E.; Frost, S.; Henry, L.; Hunt, S.; Fang, Y.; Goodman, Z.; Younossi, Z.M. Single non-invasive model to diagnose non-alcoholic fatty liver disease (NAFLD) and non-alcoholic steatohepatitis (NASH). *J. Gastroenterol. Hepatol.* **2014**, *29*, 2006–2013. [CrossRef]
16. Iqbal, U.; Perumpail, B.J.; Akhtar, D.; Kim, D.; Ahmed, A. The Epidemiology, Risk Profiling and Diagnostic Challenges of Nonalcoholic Fatty Liver Disease. *Medicines* **2019**, *6*, 41. [CrossRef]
17. Jensen, T.; Abdelmalek, M.F.; Sullivan, S.; Nadeau, K.J.; Green, M.; Roncal, C.; Nakagawa, T.; Kuwabara, M.; Sato, Y.; Kang, D.H.; et al. Fructose and sugar: A major mediator of non-alcoholic fatty liver disease. *J. Hepatol.* **2018**, *68*, 1063–1075. [CrossRef]
18. Perdomo, C.M.; Frühbeck, G.; Escalada, J. Impact of Nutritional Changes on Nonalcoholic Fatty Liver Disease. *Nutrients* **2019**, *11*, 677. [CrossRef]
19. Della Pepa, G.; Vetrani, C.; Lombardi, G.; Bozzetto, L.; Annuzzi, G.; Rivellese, A.A. Isocaloric Dietary Changes and Non-Alcoholic Fatty Liver Disease in High Cardiometabolic Risk Individuals. *Nutrients* **2017**, *26*, 1065. [CrossRef] [PubMed]
20. Valtueña, S.; Pellegrini, N.; Ardigò, D.; Del Rio, D.; Numeroso, F.; Scazzina, F.; Monti, L.; Zavaroni, I.; Brighenti, F. Dietary glycemic index and liver steatosis. *Am. J. Clin. Nutr.* **2006**, *84*, 136–142. [CrossRef] [PubMed]
21. Berná, G.; Romero-Gomez, M. The role of nutrition in non-alcoholic fatty liver disease: Pathophysiology and management. *Liver Int.* **2020**, *40*, 102–108. [CrossRef] [PubMed]
22. Zhao, H.; Yang, A.; Mao, L.; Quan, Y.; Cui, J.; Sun, Y. Association Between Dietary Fiber Intake and Non-alcoholic Fatty Liver Disease in Adults. *Front. Nutr.* **2020**, *7*, 593735. [CrossRef] [PubMed]
23. Ivancovsky-Wajcman, D.; Fliss-Isakov, N.; Salomone, F.; Webb, M.; Shibolet, O.; Kariv, R.; Zelber-Sagi, S. Dietary vitamin E and C intake is inversely associated with the severity of nonalcoholic fatty liver disease. *Dig. Liver Dis.* **2019**, *51*, 1698–1705. [CrossRef]
24. Salehi-Sahlabadi, A.; Teymoori, F.; Jabbari, M.; Momeni, A.; Mokari-Yamchi, A.; Sohouli, M.; Hekmatdoost, A. Dietary polyphenols and the odds of non-alcoholic fatty liver disease: A case-control study. *Clin. Nutr. ESPEN* **2021**, *41*, 429–435. [CrossRef] [PubMed]
25. Parry, S.A.; Hodson, L. Managing NAFLD in Type 2 Diabetes: The Effect of Lifestyle Interventions, a Narrative Review. *Adv. Ther.* **2020**, *37*, 1381–1406. [CrossRef]
26. Evert, A.B.; Dennison, M.; Gardner, C.D.; Garvey, W.T.; Lau, K.H.K.; MacLeod, J.; Mitri, J.; Pereira, R.F.; Rawlings, K.; Robinson, S.; et al. Nutrition Therapy for Adults with Diabetes or Prediabetes: A Consensus Report. *Diabetes Care* **2019**, *42*, 731–754. [CrossRef]
27. Hou, L.; Ge, L.; Wang, Q.; He, J.; Qin, T.; Cao, L.; Cao, C.; Liu, D.; Liu, X.; Yang, K. Nutritional Recommendations for Type 2 Diabetes: An International Review of 15 Guidelines. *Can. J. Diabetes* **2022**, *2*, S1499–S2671. [CrossRef]
28. Vitale, M.; Masulli, M.; Rivellese, A.A.; Bonora, E.; Babini, A.C.; Sartore, G.; Corsi, L.; Buzzetti, R.; Citro, G.; Baldassarre, M.P.A.; et al. Pasta Consumption and Connected Dietary Habits: Associations with Glucose Control, Adiposity Measures, and Cardiovascular Risk Factors in People with Type 2 Diabetes-TOSCA.IT Study. *Nutrients* **2019**, *30*, 101.
29. Vaccaro, O.; Masulli, M.; Bonora, E.; Del Prato, S.; Giorda, C.B.; Maggioni, A.P.; Mocarelli, P.; Nicolucci, A.; Rivellese, A.A.; Squatrito, S.; et al. Addition of either pioglitazone or a sulfonylurea in type 2 diabetic patients inadequately controlled with metformin alone: Impact on cardiovascular events. A randomized controlled trial. *Nutr. Metab. Cardiovasc. Dis.* **2012**, *22*, 997–1006. [CrossRef]
30. Vaccaro, O.; Masulli, M.; Nicolucci, A.; Bonora, E.; Del Prato, S.; Maggioni, A.P.; Rivellese, A.A.; Squatrito, S.; Giorda, C.B.; Sesti, G.; et al. Effects on the incidence of cardiovascular events of the addition of pioglitazone versus sulfonylureas in patients with type 2 diabetes inadequately controlled with metformin (TOSCA.IT): A randomized, multicenter trial. *Lancet Diabetes Endocrinol.* **2017**, *5*, 887–897. [CrossRef]
31. Pala, V.; Sieri, S.; Palli, D.; Salvini, S.; Berrino, F.; Bellegotti, M.; Frasca, G.; Tumino, R.; Sacerdote, C.; Fiorini, L.; et al. Diet in the Italian EPIC cohorts: Presentation of data and methodological issues. *Tumori J.* **2003**, *89*, 594–607. [CrossRef] [PubMed]
32. Pisani, P.; Faggiano, F.; Krogh, V.; Palli, D.; Vineis, P.; Berrino, F. Relative validity and reproducibility of a food frequency dietary questionnaire for use in the Italian EPIC centres. *Int. J. Epidemiol.* **1997**, *26*, S152–S160. [CrossRef]
33. Vitale, M.; Masulli, M.; Cocozza, S.; Anichini, R.; Babini, A.C.; Boemi, M.; Bonora, E.; Buzzetti, R.; Carpinteri, R.; Caselli, C.; et al. TOSCA.IT Study Group. Sex differences in food choices, adherence to dietary recommendations and plasma lipid profile in type 2 diabetes—The TOSCA.IT study. *Nutr. Metab. Cardiovasc. Dis.* **2016**, *26*, 879–885. [CrossRef]
34. Vitale, M.; Masulli, M.; Rivellese, A.A.; Bonora, E.; Cappellini, F.; Nicolucci, A.; Squatrito, S.; Antenucci, D.; Barrea, A.; Bianchi, C.; et al. TOSCA.IT Study Group. Dietary intake and major food sources of polyphenols in people with type 2 diabetes: The TOSCA.IT Study. *Eur. J. Nutr.* **2018**, *57*, 679–688. [CrossRef]
35. Salvini, S.; Parpinel, M.; Gnagnarella, P.; Maisonneuve, P.; Turrini, A. *Banca Dati di Composizione Degli Alimenti per Studi Epidemiologici in Italia*; Istituto Europeo di Oncologia: Milan, Italy, 1998.

36. Carnovale, E.; Marletta, L. *Tabella di Composizione Degli Alimenti*; Crea-Nut (ex INRAN): Rome, Italy, 2000.
37. Bhagwat, S.; Haytowitz, D.B.; Holden, J.M. USDA Database for the Flavonoid Content of Selected Foods, Release 3.1. U.S. Department of Agriculture, Agricultural Research Service. 2014; Nutrient Data Laboratory Home Page. Available online: http://www.ars.usda.gov/nutrientdata/flav (accessed on 16 September 2022).
38. Rothwell, J.A.; Pérez-Jiménez, J.; Neveu, V.; Medina-Ramon, A.; M'Hiri, N.; Garcia Lobato, P.; Manach, C.; Knox, K.; Eisner, R.; Wishart, D.; et al. Phenol-Explorer 3.0: A Major Update of the Phenol-Explorer Database to Incorporate Data on the Effects of Food Processing on Polyphenol Content. Database. 2013. Available online: https://doi.org/10.1093/database/bat070 (accessed on 26 September 2022). [CrossRef]
39. Buckland, G.; González, C.A.; Agudo, A.; Vilardell, M.; Berenguer, A.; Amiano, P.; Ardanaz, E.; Arriola, L.; Barricarte, A.; Basterretxea, M.; et al. Adherence to the Mediterranean diet and risk of coronary heart disease in the Spanish EPIC Cohort Study. *Am. J. Epidemiol.* **2009**, *170*, 1518–1529. [CrossRef]
40. Vitale, M.; Masulli, M.; Calabrese, I.; Rivellese, A.A.; Bonora, E.; Signorini, S.; Perriello, G.; Squatrito, S.; Buzzetti, R.; Sartore, G.; et al. TOSCA.IT Study Group. Impact of a Mediterranean Dietary Pattern and Its Components on Cardiovascular Risk Factors, Glucose Control, and Body Weight in People with Type 2 Diabetes: A Real-Life Study. *Nutrients* **2018**, *10*, 1067. [CrossRef]
41. Yki-Järvinen, H.; Luukkonen, P.K.; Hodson, L.; Moore, J.B. Dietary carbohydrates and fats in nonalcoholic fatty liver disease. *Nat. Rev. Gastroenterol. Hepatol.* **2021**, *18*, 770–786. [CrossRef]
42. Fan, J.G.; Cao, H.X. Role of diet and nutritional management in non-alcoholic fatty liver disease. *J. Gastroenterol. Hepatol.* **2013**, *28*, 81–87. [CrossRef]
43. Zelber-Sagi, S.; Ratziu, V.; Oren, R. Nutrition and physical activity in NAFLD: An overview of the epidemiological evidence. *World J. Gastroenterol.* **2011**, *17*, 3377–3389. [CrossRef] [PubMed]
44. He, K.; Li, Y.; Guo, X.; Zhong, L.; Tang, S. Food groups and the likelihood of non-alcoholic fatty liver disease: A systematic review and meta-analysis. *Br. J. Nutr.* **2020**, *124*, 1–13. [CrossRef] [PubMed]
45. Chung, M.; Ma, J.; Patel, K.; Berger, S.; Lau, J.; Lichtenstein, A.H. Fructose, high-fructose corn syrup, sucrose, and nonalcoholic fatty liver disease or indexes of liver health: A systematic review and meta-analysis. *Am. J. Clin. Nutr.* **2014**, *100*, 833–849. [CrossRef] [PubMed]
46. Bayram, H.M.; Majoo, F.M.; Ozturkcan, A. Polyphenols in the prevention and treatment of non-alcoholic fatty liver disease: An update of preclinical and clinical studies. *Clin. Nutr. ESPEN.* **2021**, *44*, 1–14. [CrossRef] [PubMed]
47. Abenavoli, L.; Larussa, T.; Corea, A.; Procopio, A.C.; Boccuto, L.; Dallio, M.; Federico, A.; Luzza, F. Dietary Polyphenols and Non-Alcoholic Fatty Liver Disease. *Nutrients* **2021**, *13*, 494. [CrossRef] [PubMed]
48. Abe, R.A.M.; Masroor, A.; Khorochkov, A.; Prieto, J.; Singh, K.B.; Nnadozie, M.C.; Abdal, M.; Shrestha, N.; Mohammed, L. The Role of Vitamins in Non-Alcoholic Fatty Liver Disease: A Systematic Review. *Cureus* **2021**, *13*, e16855. [CrossRef]
49. Rodriguez-Ramiro, I.; Vauzour, D.; Minihane, A.M. Polyphenols and non-alcoholic fatty liver disease: Impact and mechanisms. *Proc. Nutr. Soc.* **2016**, *75*, 47–60. [CrossRef]
50. Pérez-Montes de Oca, A.; Julián, M.T.; Ramos, A.; Puig-Domingo, M.; Alonso, N. Microbiota, Fiber, and NAFLD: Is There Any Connection? *Nutrients* **2020**, *12*, 3100. [CrossRef]
51. Bozzetto, L.; Annuzzi, G.; Ragucci, M.; Di Donato, O.; Della Pepa, G.; Della Corte, G.; Griffo, E.; Anniballi, G.; Giacco, A.; Mancini, M.; et al. Insulin resistance, postprandial GLP-1 and adaptive immunity are the main predictors of NAFLD in a homogeneous population at high cardiovascular risk. *Nutr. Metab. Cardiovasc. Dis.* **2016**, *26*, 623–629. [CrossRef]
52. Zelber-Sagi, S.; Ivancovsky-Wajcman, D.; Fliss Isakov, N.; Webb, M.; Orenstein, D.; Shibolet, O.; Kariv, R. High red and processed meat consumption is associated with non-alcoholic fatty liver disease and insulin resistance. *J. Hepatol.* **2018**, *68*, 1239–1246. [CrossRef]
53. Hashemian, M.; Poustchi, H.; Merat, S.; Abnet, C.; Malekzadeh, R.; Etemadi, A. Red Meat Consumption and Risk of Nonalcoholic Fatty Liver Disease in a Population with Low Red Meat Consumption. *Curr. Dev. Nutr.* **2020**, *29*, 1413. [CrossRef]
54. Akhlaghi, M.; Ghasemi-Nasab, M.; Riasatian, M. Mediterranean diet for patients with non-alcoholic fatty liver disease, a systematic review and meta-analysis of observational and clinical investigations. *J. Diabetes Metab. Disord.* **2020**, *19*, 575–584. [CrossRef] [PubMed]
55. Castera, L.; Friedrich-Rust, M.; Loomba, R. Noninvasive Assessment of Liver Disease in Patients with Nonalcoholic Fatty Liver Disease. *Gastroenterology* **2019**, *156*, 1264–1281. [CrossRef] [PubMed]

Article

Dietary Isorhamnetin Intake Is Associated with Lower Blood Pressure in Coronary Artery Disease Patients

Joanna Popiolek-Kalisz [1,2,*], Piotr Blaszczak [2] and Emilia Fornal [3]

1 Clinical Dietetics Unit, Department of Bioanalytics, Medical University of Lublin, ul. Chodzki 7, 20-093 Lublin, Poland
2 Department of Cardiology, Cardinal Wyszynski Hospital in Lublin, al. Krasnicka 100, 20-718 Lublin, Poland
3 Department of Bioanalytics, Medical University of Lublin, ul. Jaczewskiego 8b, 20-090 Lublin, Poland
* Correspondence: joannapopiolekkalisz@umlub.pl

Abstract: Background: Recent studies suggest the positive role of flavonols on blood pressure (BP) values, although there are not many conducted on humans. The aim of this study was to examine the relationship between flavonol intake and their main sources of consumption, and systolic (SBP) and diastolic (DBP) BP values in coronary artery disease (CAD) patients. Methods and results: forty CAD patients completed a food-frequency questionnaire dedicated to flavonol-intake assessment. The analysis revealed significant correlation between isorhamnetin intake and SBP values—absolute (R: −0.36; 95% CI: −0.602 to −0.052; $p = 0.02$), and related to body mass (R: −0.38; 95% CI: −0.617 to −0.076; $p = 0.02$. This effect was observed in male participants (R: −0.65; 95% CI: −0.844 to −0.302; $p = 0.001$ and R: −0.63; 95% CI: −0.837 to −0.280; $p = 0.002$ respectively), but not in female patients. The main contributors were onions, tomatoes, blueberries, apples, tea, coffee and wine. White onion (R: −0.39; 95% CI: −0.624 to −0.088; $p = 0.01$) consumption was inversely correlated with SBP, and tomato consumption (R: −0.33; 95% CI: −0.581 to −0.020; $p = 0.04$) with DBP. The comparison between patients with BP < 140 mmHg and ≥140 mmHg revealed significant differences in white onion ($p = 0.01$) and blueberry ($p = 0.04$) intake. Conclusions: This study revealed the relationship between long-term dietary isorhamnetin intake and SBP values. The analysis of specific food intake showed that onion, tomato and blueberry consumption could impact BP values. This may suggest that a dietary approach which includes a higher intake of isorhamnetin-rich products could possibly result in BP lowering in CAD patients.

Keywords: flavonols; quercetin; hypertension; blood pressure; isorhamnetin

Citation: Popiolek-Kalisz, J.; Blaszczak, P.; Fornal, E. Dietary Isorhamnetin Intake Is Associated with Lower Blood Pressure in Coronary Artery Disease Patients. *Nutrients* **2022**, *14*, 4586. https://doi.org/10.3390/nu14214586

Academic Editor: Giuseppe Della Pepa

Received: 6 October 2022
Accepted: 28 October 2022
Published: 1 November 2022

Publisher's Note: MDPI stays neutral with regard to jurisdictional claims in published maps and institutional affiliations.

Copyright: © 2022 by the authors. Licensee MDPI, Basel, Switzerland. This article is an open access article distributed under the terms and conditions of the Creative Commons Attribution (CC BY) license (https://creativecommons.org/licenses/by/4.0/).

1. Introduction

Elevated blood pressure (BP) is the leading cardiovascular disease (CVD) in Poland and it is present in 9.94 million adults, which is about 26% of total Polish population [1]. The overall prevalence of hypertension in adults is around 30–45% [2]. It is also a recognized cardiovascular risk factor, which is why finding a pattern leading to this condition is the main target of primary and secondary prevention. Apart from pharmacological treatment, which is introduced after a hypertension diagnosis, lifestyle changes including dietary approach are essential to prevent this condition, and as the first-line treatment [3]. The European Society of Cardiology recommendations for hypertension management include a diet rich in vegetables and fruit, although they are not very precise [3]. Vegetables and fruits are the sources of flavonoids which are investigated in varying contexts of human health, due to their antioxidative properties. Flavonols are the group of flavonoids distinguished by their chemical structure including a 3-hydroxyflavone backbone. They differ in the presence and position of hydroxyl and methyl groups. The main flavonols are quercetin, kaempferol, isorhamnetin and myricetin, although there is a large group of flavonols which are less abundant in the everyday diet, e.g., morin, galangin, fisetin, kaempferide, azaleatin, natsudaidain, pachypodol and rhamnazin [4–6]. The bioactivity of each compound depends on the number and type of functional groups. Quercetin, kaempferol and myricetin differ in the number of hydroxyl groups, while isorhamnetin is O-methylated in the R3 position, compared to quercetin.

The products particularly rich in flavonols are onions, tea, and apples, although kale, lettuce, tomatoes, broccoli, grapes, berries and red wine are also known to be flavonol-rich [6–8]. The main dietary contributors for quercetin intake are tomatoes, kale, apples and tea; for an intake of kaempferol, kale, beans, tea, spinach, and broccoli; for isorhamnetin, pears, olive oil, wine, and tomato sauce; for myricetin intake, tea, wine, kale, oranges, and tomatoes [9].

The most investigated flavonol is quercetin. The interventional studies in humans suggest the impact of its supplementation in BP regulation [10–13]. The other flavonols have not been the subject of interventional studies in humans yet, although the studies on animal models also suggest a positive role for isorhamnetin supplementation in hypertension management [14]. What is more, its potential role as a cardioprotective, neuroprotective, anti-tumor and anti-obesity agent was also suggested in in vitro and animal model studies [15–18]. Nonetheless, the results from the only observational human study which investigated the relationship between dietary-antioxidant habitual intake and hypertension are not consistent with this, as there was no observed correlation between flavonol (quercetin, kaempferol, isorhamnetin and myricetin) intake and hypertension incidence [7]. Nonetheless, it is worth noting that general hypertension diagnosis is based on crossing the limit of 140 mmHg for systolic BP (SBP) and/or 90 mmHg for diastolic BP (DBP) [3]. There has not yet been any study investigating the linear relationship between flavonol intake and BP values in coronary artery disease (CAD).

The consumption of apples, which are a main source of flavonols, is generally advised in terms of health benefits ("an apple a day keeps a doctor away"); however, there have not been any studies which have analyzed the impact of apple consumption on BP values [19]. On the other hand, patients are often discouraged from drinking coffee, which is also a good source of flavonols, due to its potential negative impact on BP values, even though the recent studies do not confirm this [20].

The aim of this study was to analyze the impact of long-term dietary intake of the selected flavonols (quercetin, kaempferol, isorhamnetin and myricetin) and their main dietary sources, on levels of SBP and DBP among patients with CAD. Additionally, the impact of long-term consumption of apples and coffee, which are sources of flavonols, on BP values, was also investigated.

2. Materials and Methods

Forty adult patients hospitalized between March and July 2022 due to CAD were enrolled in this study. Inclusion criteria were: (1) CAD diagnosis (2) age $\geq$ 18 years (3) written consent (4) mental condition that enabled a one-year retrospective dietary interview. The food-frequency questionnaire dedicated to specific flavonol one-year-intake assessment was administered to the patients [21]. On the basis of this, the mean daily intake for quercetin, kaempferol, isorhamnetin, myricetin and total flavonols was calculated for each patient. The information about mean daily intake of flavonol sources was also derived from the questionnaire.

Patient weight was measured to 0.05 kg accuracy using the WTL-150A scale (Lubelskie Fabryki Wag), by a trained professional. The patient was allowed to wear only underwear for this measurement. BP was measured by a trained professional to 1 mmHg accuracy with the Omron M3 monitor (Omron Healthcare). The measurement was performed according to the European Society of Cardiology recommendations [22].

The study was approved by the local Bioethics Committee of the Medical University of Lublin (consent no. KE-0254/9/01/2022). The study was conducted in line with the directives of the Declaration of Helsinki on Ethical Principles for Medical Research. All participants signed a written consent agreement.

Statistical analyses were performed with the RStudio software v. 4.2.0. The normality of the distribution of each parameter was checked by the Shapiro–Wilk test. The variables were presented as means (SD). Pearson correlation was used to analyze the association between selected flavonol mean daily-intake and SBP or DBP, and between selected products mean daily-intake and SBP or DBP. The cut-off points used for correlation coefficient were as follows: <0.20 as low, 0.20–0.49 as moderate and ≥0.50 as high correlation. A p value below 0.05 was considered significant.

The patients were also divided into two groups due to SBP value—below 140 mmHg and 140 mmHg or higher. The differences in selected flavonol mean daily-intake and selected mean daily products between the groups were investigated by the Mann–Whitney test. A p value below 0.05 was considered significant.

3. Results

A total of 40 patients (21 men and 19 women) were enrolled in the study. The mean age was 68 (±9) years and mean weight was 80.40 (±10.30) kg. The mean daily total flavonol intake was 62.64 (±33.98) mg/day, and for specific flavonols: 29.77 (±22.18) mg/day for quercetin, 14.86 (±8.56) mg/day for kaempferol, 2.46 (±2.02) mg/day for isorhamnetin and 5.55 (±4.16) mg/day for myricetin. When the values were referred to body mass, the mean daily intake was 0.80 (±0.45) mg/kg for total flavonols, 0.51 (±0.29) mg/kg for quercetin, 0.19 (±0.11) mg/kg for kaempferol, 0.03 (±0.02) mg/kg for isorhamnetin and 0.07 (±0.05) mg/kg for myricetin. The mean measured SBP was 134.23 (±25.49) mmHg and DBP was 73.50 (±11.29) mmHg. The main contributors to flavonol intake were onions (white and red), tomatoes, blueberries, apples, tea, coffee and wine.

The results revealed a significant moderate correlation between daily isorhamnetin intake and SBP values. The relation was present in both absolute daily intake (R: −0.36; 95% CI: −0.602 to −0.052; p = 0.02) and daily intake related to body mass (R: −0.38; 95% CI −0.617 to −0.076; p = 0.02). The detailed results for all analyzed flavonols are presented in Table 1.

Table 1. The correlation between selected flavonol daily intake and systolic and diastolic blood pressure.

	Systolic Blood Pressure			Diastolic Blood Pressure		
	R	95% CI	p	R	95% CI	p
Quercetin daily intake	−0.23	−0.506; 0.087	0.15	0.05	−0.264; 0.357	0.75
Kaempferol daily intake	0.02	−0.292; 0.331	0.89	0.19	−0.125; 0.477	0.23
Isorhamnetin daily intake	−0.36	−0.602; −0.052	0.02	0.05	−0.263; 0.359	0.74
Myricetin daily intake	−0.08	−0.379; 0.241	0.64	0.09	−0.230; 0.388	0.59
Total flavonol daily intake	−0.18	−0.462; 0.143	0.28	0.10	−0.222; 0.396	0.55
Quercetin daily intake/body mass	−0.28	−0.542; 0.037	0.08	0.04	−0.272; 0.350	0.79
Kaempferol daily intake/body mass	−0.05	−0.354; 0.268	0.77	0.18	−0.138; 0.466	0.26
Isorhamnetin daily intake/body mass	−0.38	−0.617; −0.076	0.02	0.07	−0.250; 0.370	0.68
Myricetin daily intake/body mass	−0.13	−0.426; 0.187	0.41	0.07	−0.247; 0.374	0.67
Total flavonol daily intake/body mass	−0.23	−0.503; 0.091	0.16	0.09	−0.232; 0.387	0.60

The analysis of the male and female subgroup revealed that this effect was observed only in male participants (R: −0.65; 95% CI: −0.844 to −0.302; p = 0.001 for absolute isorhamnetin intake and R: −0.63; 95% CI: −0.837 to −0.280; p = 0.002 for related-to-body-mass isorhamnetin intake). The detailed results are presented in Tables 2 and 3.

Table 2. The correlation between selected flavonols daily intake and systolic and diastolic blood pressure in male participants.

	Systolic Blood Pressure			Diastolic Blood Pressure		
	R	95% CI	*p*	R	95% CI	*p*
Quercetin daily intake	−0.32	−0.661; 0.128	0.16	−0.14	−0.542; 0.307	0.53
Kaempferol daily intake	0.11	−0.340; 0.515	0.64	0.01	−0.422; 0.441	0.96
Isorhamnetin daily intake	−0.65	−0.844; −0.302	0.001	−0.12	−0.515; 0.340	0.64
Myricetin daily intake	−0.06	−0.478; 0.383	0.80	−0.11	−0.518; 0.336	0.63
Total flavonol daily intake	−0.27	−0.626; 0.187	0.24	−0.13	−0.530; 0.323	0.58
Quercetin daily intake/body mass	−0.27	−0.628; 0.183	0.24	−0.08	−0.497; 0.361	0.72
Kaempferol daily intake/body mass	0.16	−0.292; 0.554	0.49	0.11	−0.338; 0.517	0.63
Isorhamnetin daily intake/body mass	−0.63	−0.837; −0.280	0.002	−0.07	−0.491; 0.369	0.75
Myricetin daily intake/body mass	−0.03	−0.457; 0.405	0.89	−0.08	−0.491; 0.369	0.75
Total flavonol daily intake/body mass	−0.21	−0.585; 0.248	0.37	−0.05	−0.470; 0.392	0.84

Table 3. The correlation between selected flavonol daily intake and systolic and diastolic blood pressure in female participants.

	Systolic Blood Pressure			Diastolic Blood Pressure		
	R	95% CI	*p*	R	95% CI	*p*
Quercetin daily intake	−0.17	−0.580; 0.308	0.49	0.30	−0.180; 0.663	0.21
Kaempferol daily intake	−0.09	−0.521; 0.381	0.72	0.34	−0.136; 0.687	0.16
Isorhamnetin daily intake	0.01	−0.447; 0.461	0.97	0.34	−0.130; 0.691	0.15
Myricetin daily intake	−0.14	−0.556; 0.339	0.58	0.37	−0.101; 0.706	0.12
Total flavonol daily intake	−0.14	−0.556; 0.339	0.58	0.32	−0.157; 0676	0.18
Quercetin daily intake / body mass	−0.27	−0.643; 0.213	0.27	0.27	−0.206; 0.647	0.26
Kaempferol daily intake / body mass	−0.19	−0.591; 0.292	0.44	0.30	−0.179; 0.664	0.21
Isorhamnetin daily intake / body mass	−0.06	−0.502; −0.404	0.80	0.37	−0.105; 0.703	0.12
Myricetin daily intake / body mass	−0.23	−0.618; 0.252	0.35	0.32	−0.161; 0.674	0.19
Total flavonol daily intake / body mass	−0.24	−0.625; 0.242	0.33	0.29	−0.192; 0.656	0.23

The analysis of the main flavonol sources mean daily intake showed that onion (R: −0.38; 95% CI −0.616 to −0.074; $p = 0.02$) and white onion (R: −0.39; 95% CI: −0.624 to −0.088; $p = 0.01$) intake is correlated with SBP values, while tomato intake (R: −0.33; 95% CI: −0.581 to −0.020; $p = 0.04$) is correlated with DBP values. The detailed results for all analyzed products are presented in the Table 4.

Table 4. The correlation between selected product daily intake and systolic and diastolic blood pressure.

	Systolic Blood Pressure			Diastolic Blood Pressure		
	p	95% CI	R	*p*	95% CI	R
White onion	0.01	−0.624; −0.088	−0.39	0.87	−0.335; 0.288	−0.03
Red onion	0.15	0.508; 0.084	−0.23	0.34	−0.166; 0.444	0.15
Onion (total)	0.02	−0.616; −0.073	−0.38	0.76	−0.265; 0.357	0.05
Tomatoes	0.31	−0.454; 0.153	−0.17	0.04	−0.581; −0.020	−0.33
Blueberry	0.21	−0.483; 0.116	−0.20	0.79	−0.349; 0.273	−0.04
Apples	0.39	−0.431; 0.181	−0.14	0.68	−0.371; 0.249	−0.07
Black tea	0.58	−0.228; 0.391	0.09	0.22	−0.121; 0.480	0.20
Green tea	0.57	−0.393; 0.225	−0.09	0.25	−0.132; 0.472	0.19
Coffee	0.97	−0.306; 0.317	0.01	0.96	−0.318; 0.305	−0.01
Wine	0.89	−0.337; 0.294	−0.02	0.52	−0.215; 0.409	0.11

The subgroup analysis revealed significant differences in isorhamnetin intake related to body mass ($p = 0.048$), white onions ($p = 0.01$) and blueberries ($p = 0.04$) among the patients with normal BP (<140 mmHg) and elevated BP (≥140 mmHg). The detailed results for all compounds and food are presented in the Table 5 and Figure 1. The additional analysis between the patients consuming less than 1 apple a day and ≥1 apple daily did not show any significant differences in terms of SBP (p = 0.55) or DBP (p = 0.95). A similar observation was made for coffee consumption (p = 0.64 for SBP and p = 0.43 for DBP).

Table 5. Differences in flavonol and selected product daily intake in patients with normal systolic blood pressure (<140 mmHg) and elevated systolic blood pressure (≥140 mmHg).

	Systolic Blood Pressure				
	<140 mmHg		**≥140 mmHg**		
	Mean	**SD**	**Mean**	**SD**	***p***
Quercetin [mg/day]	42.02	±24.81	36.73	±17.24	0.61
Kaempferol [mg/day]	13.91	±8.66	16.14	±8.24	0.39
Isorhamnetin [mg/day]	2.88	±2.20	1.90	±1.53	0.08
Myricetin [mg/day]	5.66	±4.80	5.39	±3.02	0.59
Total flavonols [mg/day]	72.09	±39.77	67.57	±30.94	0.94
Quercetin [mg/kg*day]	0.55	±0.33	0.45	±0.22	0.55
Kaempferol [mg/kg*day]	0.18	±0.12	0.20	±0.10	0.52
Isorhamnetin [mg/kg*day]	0.04	±0.03	0.02	±0.02	0.048
Myricetin [mg/kg*day]	0.07	±0.06	0.07	±0.04	0.94
Total flavonols [mg/kg*day]	0.94	± 0.54	0.83	±0.40	0.68
White onion [portion/day]	0.31	±0.25	0.17	±0.25	0.01
Red onion [portion/day]	0.09	±0.21	0.07	±0.06	0.51
Tomatoes [portion/day]	0.58	±0.82	0.34	±0.27	0.64
Blueberries [portion/day]	0.24	±0.31	0.06	±0.06	0.04
Apples [portion/day]	0.68	±0.51	0.56	±0.47	0.65
Black tea [portion/day]	1.26	±1.22	2.27	±2.03	0.10
Green tea [portion/day]	0.49	±0.77	0.45	±1.08	0.44
Coffee [portion/day]	0.76	±0.88	0.39	±0.47	0.37
Wine [portion/day]	0.05	±0.10	0.11	±0.29	0.82

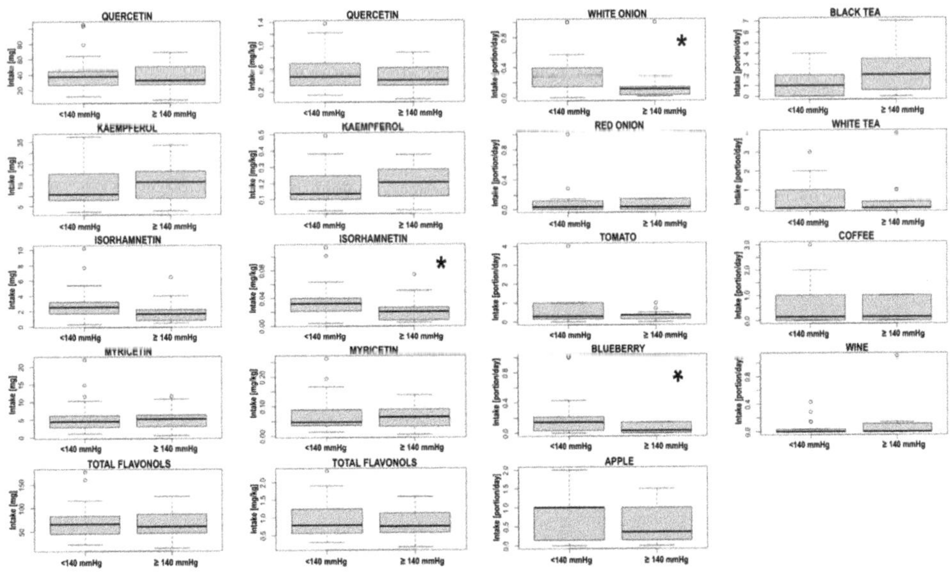

Figure 1. Box-plots presenting differences in flavonol intake and their main dietary source consumption among patients with systolic blood pressure <140 mmHg and ≥140 mmHg (significant differences are marked with *).

4. Discussion

Flavonols are the subgroup of flavonoids which share a 3-hydroxyflavone backbone. Individual flavonols differ in their chemical structure (i.e., presence and position of hydroxyl and methyl groups) which impacts their bioactivity [23,24]. The most investigated flavonol is quercetin, and the studies in humans suggest the positive role of its supplementation in BP regulation [10–13]. The other flavonols are not as widely analyzed in this context. Most of the interventional studies about isorhamnetin impact on BP come from animal models, where isorhamnetin restored vasodilatation in hypertensive rats [14]. The mechanisms of this effect are still under investigation. However, the role of modification of protein kinases (C and Rho) activity, cytosolic H_2O_2 production or angiotensin-converting enzyme inhibition potential are suggested by other authors [14,25–27]. On the other hand, in the only observational human study which analyzed the impact of dietary antioxidant intake on hypertension, there was no correlation between general habitual dietary flavonoid intake and a reduction in incidents of hypertension [7]. A similar observation was also made in this study for specific flavonols (quercetin, kaempferol, isorhamnetin and myricetin) [7]. It is worth noting that the study was conducted on a very large population (a total of 156 957 participants from the Nurses' Health Study and Health Professionals follow-up study); nonetheless, it was based on patients' self-assessment and self-reporting, and analyzed the impact of the above-mentioned dietary agents only on hypertension incidence, without taking exact BP-value measurements into consideration [7]. The flavonol intake in that study was also calculated on the basis of a general semiquantitative food-frequency questionnaire, and not by the dedicated tool [7]. That is why it could possibly omit some of the subtle relationships between isorhamnetin intake and BP values shown in this study. The products particularly rich in isorhamnetin are onions (white and red), kale, asparagus, elderberry, dill and parsley [6,8].

The presented results revealed also significant differences between male and female participants in terms of the size of the effect of isorhamnetin consumption. These effects were observed only in the male subgroup, while the correlation was not significant in the female subgroup. This is an interesting observation, as in the study by Knekt et al. the effect of flavonoid intake on general CAD mortality was observed in women, but not in men [28]. It is worth noting that hypertension is a CAD risk factor, but in the study by Knekt et al. the direct BP values were not taken into consideration. However, in the studies which analyzed the impact of other flavonol intake (quercetin) directly on hypertension, it was shown that quercetin supplementation results in antihypertensive effects in men [29]. Nonetheless, the present study is the first study which has investigated the antihypertensive potential of dietary isorhamnetin in humans.

The results showed that the main contributors to flavonol intake were onions, tomatoes among the vegetables; blueberries and apples among the fruit; and tea, coffee and wine among the beverages, which matches the observations made in the Zutphen Elderly study [30]. The analysis showed that among these selected products, only onions and tomato mean daily-intakes were significantly correlated with SBP and DBP, respectively. The subgroup analysis confirmed this observation, as the patients with elevated SBP were characterized by significantly lower white onion consumption. Onion extract was already proven to have potential in BP-lowering, although the doses used in the study by Brull et al. and Kalus et al. were higher than reachable in an everyday diet [31,32]. The impact of onion and apple intake on cardiovascular mortality was shown in the study by Knekt et al. and Hertog et al. [28,30]. The subgroup analysis showed that patients with elevated and normal BP also significantly differed in terms of blueberry consumption. This relationship was not presented for berries in the above-mentioned study; however, it could be explained that in the presented study berries were divided into species, and only blueberry intake was taken into consideration, in contrast with the other studies [28]. Even though the studies did not analyze the direct values of BP, it is worth noting that hypertension is a cardiovascular risk factor. The relation between the intake of food rich in flavonoids such as onions, apples or tea, and cardiovascular risk factors (including BP) was also shown in French women, although the authors of the SU.VI.MAX study did not reveal the exact correlations [33]. The important factor could be also the preparation of the meals, including the products mentioned, as boiling could decrease the antihypertensive potential in onions [34].

Apples consumption is popular, due to the beneficial role for health. The term "an apple a day keeps a doctor away" was examined in the course of this study, and it revealed that patients who consume one apple a day or more do not have significantly lower SBP or DBP, compared to the patients with lower apple-consumption. This observation might be caused by the lower border for the minimal apple-consumption impact, as the study investigating the impact of fresh fruit intake and acute coronary syndrome proved that the level of consumption of 25 g/day reduces the risk of acute coronary syndromes [19]. What is more, coffee consumption, which is suggested to elevate BP, did not present such properties in CAD patients. The patients who drink one coffee a day or more did not have significantly higher SBP or DBP compared to the patients who drink less than one coffee a day. This observation matches the results from other studies regarding CVD risk and coffee consumption [20].

On the basis of the presented results, we can suggest that incorporating products such as onion, tomatoes and blueberries into the everyday diet could be possibly beneficial in terms of BP values. Nonetheless, a longer observation on a larger population, or ideally, a controlled prospective study is needed to support this.

In our observational study, the mean daily intake of quercetin was much lower (29.77 [$\pm$22.18] mg/day) than the supplementation doses used in randomized controlled studies in humans (50 mg to 730 mg/day), so this may explain why the results from this observation do not match the results from the meta-analysis, which showed that quercetin supplementation could decrease BP values [12]. It is also worth noting that bioavailability from an artificial supplement can differ from that from a dietary source.

Apart from the mentioned study, there are no other available observational or interventional studies focused on the impact of other flavonols (kaempferol and myricetin) on BP level.

5. Conclusions

This study revealed the relationship between long-term dietary isorhamnetin consumption and SBP values in male patients. The correlation was not proved for other flavonols or for DBP. The analysis of specific foods showed that onion, tomato and blueberry intake could impact BP values. This may suggest that a dietary approach which includes a higher intake of products rich in this compound could possibly result in BP lowering.

Author Contributions: Conceptualization, J.P.-K.; data curation, J.P.-K.; funding acquisition, E.F.; investigation, J.P.-K.; methodology, J.P.-K.; resources, J.P.-K. and P.B.; visualization, J.P.-K.; writing—original draft, J.P.-K.; writing—review and editing, J.P.-K., P.B. and E.F. All authors edited and approved the final version of the manuscript.

Funding: This study was supported by the Ministry of Education and Science in Poland within the statutory activity of the Medical University of Lublin (DS 472/2022).

Institutional Review Board Statement: The study was conducted in accordance with the Declaration of Helsinki, and approved by the local Ethics Committee of the Medical University of Lublin (consent no. KE-0254/9/01/2022).

Informed Consent Statement: Informed consent was obtained from all patients involved in the study.

Data Availability Statement: The data that support the findings of this study are available from the corresponding author upon reasonable request.

Acknowledgments: The authors would like to thank Grzegorz Kalisz for the support in the statistical analysis.

Conflicts of Interest: The authors declare no conflict of interest.

References

1. Narodowy Fundusz Zdrowia. *NFZ o Zdrowiu. Nadciśnienie Tętnicze*; Narodowy Fundusz Zdrowia: Warsaw, Poland, 2019.
2. Chow, C.K.; Teo, K.K.; Rangarajan, S.; Islam, S.; Gupta, R.; Avezum, A.; Bahonar, A.; Chifamba, J.; Dagenais, G.; Diaz, R.; et al. Prevalence, awareness, treatment, and control of hypertension in rural and urban communities in high-, middle-, and low-income countries. *JAMA* **2013**, *310*, 959–968. [CrossRef]
3. Williams, B.; Mancia, G.; Spiering, W.; Agabiti Rosei, E.; Azizi, M.; Burnier, M.; Clement, D.; Coca, A.; De Simone, G.; Dominiczak, A.; et al. 2018 Practice guidelines for the management of arterial hypertension of the European Society of Hypertension (ESH) and the European Society of Cardiology (ESC). *Blood Press.* **2018**, *27*, 314–340. [CrossRef] [PubMed]
4. Zamora-Ros, R.; Andres-Lacueva, C.; Lamuela-Raventós, R.M.; Berenguer, T.; Jakszyn, P.; Barricarte, A.; Ardanaz, E.; Amiano, P.; Dorronsoro, M.; Larrañaga, N.; et al. Estimation of Dietary Sources and Flavonoid Intake in a Spanish Adult Population (EPIC-Spain). *J. Am. Diet. Assoc.* **2010**, *110*, 390–398. [CrossRef] [PubMed]
5. Sampson, L.; Rimm, E.; Hollma, P.C.H.; de Vries, J.H.M.; Katan, M.B. Flavonol and Flavone Intakes in US Health Professionals. *J. Am. Diet. Assoc.* **2002**, *102*, 1414–1420. [CrossRef]
6. Zamora-Ros, R.; Knaze, V.; Luján-Barroso, L.; Slimani, N.; Romieu, I.; Fedirko, V.; Santucci de Magistris, M.; Ericson, U.; Amiano, P.; Trichopoulou, A.; et al. Estimated dietary intakes of flavonols, flavanones and flavones in the European Prospective Investigation into Cancer and Nutrition (EPIC) 24 hour dietary recall cohort. *Br. J. Nutr.* **2011**, *106*, 1915–1925. [CrossRef]
7. Cassidy, A.; O'Reilly, É.J.; Kay, C.; Sampson, L.; Franz, M.; Forman, J.; Curhan, G.; Rimm, E.B. Habitual intake of flavonoid subclasses and incident hypertension in adults. *Am. J. Clin. Nutr.* **2011**, *93*, 338–347. [CrossRef]
8. Panche, A.N.; Diwan, A.D.; Chandra, S.R. Flavonoids: An overview. *J. Nutr. Sci.* **2016**, *5*, e47. [CrossRef]
9. Holland, T.M.; Agarwal, P.; Wang, Y.; Leurgans, S.E.; Bennett, D.A.; Booth, S.L.; Morris, M.C. Dietary flavonols and risk of Alzheimer dementia. *Neurology* **2020**, *94*, e1749–e1756. [CrossRef]
10. Conquer, J.A.; Maiani, G.; Azzini, E.; Raguzzini, A.; Holub, B.J. Supplementation with Quercetin Markedly Increases Plasma Quercetin Concentration without Effect on Selected Risk Factors for Heart Disease in Healthy Subjects. *J. Nutr.* **1998**, *128*, 593–597. [CrossRef]
11. Edwards, R.L.; Lyon, T.; Litwin, S.E.; Rabovsky, A.; Symons, J.D.; Jalili, T. Quercetin Reduces Blood Pressure in Hypertensive Subjects. *J. Nutr.* **2007**, *137*, 2405–2411. [CrossRef]
12. Popiolek-Kalisz, J.; Fornal, E. The Effects of Quercetin Supplementation on Blood Pressure—Meta-Analysis. *Curr. Probl. Cardiol.* **2022**, *47*, 101350. [CrossRef] [PubMed]
13. Popiolek-Kalisz, J.; Fornal, E. The Impact of Flavonols on Cardiovascular Risk. *Nutrients* **2022**, *14*, 1973. [CrossRef] [PubMed]
14. Ibarra, M.; Moreno, L.; Vera, R.; Cogolludo, A.; Duarte, J.; Tamargo, J.; Perez-Vizcaino, F. Effects of the Flavonoid Quercetin and its Methylated Metabolite Isorhamnetin in Isolated Arteries from Spontaneously Hypertensive Rats. *Planta Med.* **2003**, *69*, 995–1000. [CrossRef] [PubMed]
15. Gao, L.; Yao, R.; Liu, Y.; Wang, Z.; Huang, Z.; Du, B.; Zhang, D.; Wu, L.; Xiao, L.; Zhang, Y. Isorhamnetin protects against cardiac hypertrophy through blocking PI3K–AKT pathway. *Mol. Cell. Biochem.* **2017**, *429*, 167–177. [CrossRef]
16. Gong, G.; Guan, Y.Y.; Zhang, Z.L.; Rahman, K.; Wang, S.J.; Zhou, S.; Luan, X.; Zhang, H. Isorhamnetin: A review of pharmacological effects. *Biomed. Pharmacother.* **2020**, *128*, 110301. [CrossRef]
17. Rodríguez-Rodríguez, C.; Torres, N.; Gutiérrez-Uribe, J.A.; Noriega, L.G.; Torre-Villalvazo, I.; Leal-Díaz, A.M.; Antunes-Ricardo, M.; Márquez-Mota, C.; Ordaz, G.; Chavez-Santoscoy, R.A.; et al. The effect of isorhamnetin glycosides extracted from Opuntia ficus-indica in a mouse model of diet induced obesity. *Food Funct.* **2015**, *6*, 805–815. [CrossRef]
18. Antunes-Ricardo, M.; Guardado-Félix, D.; Rocha-Pizaña, M.R.; Garza-Martínez, J.; Acevedo-Pacheco, L.; Gutiérrez-Uribe, J.A.; Villela-Castrejón, J.; López-Pacheco, F.; Serna-Saldívar, S.O. Opuntia ficus-indica Extract and Isorhamnetin-3-O-Glucosyl-Rhamnoside Diminish Tumor Growth of Colon Cancer Cells Xenografted in Immune-Suppressed Mice through the Activation of Apoptosis Intrinsic Pathway. *Plant Foods Hum. Nutr.* **2021**, *76*, 434–441. [CrossRef]
19. Hansen, L.; Dragsted, L.O.; Olsen, A.; Christensen, J.; Tjønneland, A.; Schmidt, E.B.; Overvad, K. Fruit and vegetable intake and risk of acute coronary syndrome. *Br. J. Nutr.* **2010**, *104*, 248–255. [CrossRef]
20. Ding, M.; Bhupathiraju, S.N.; Satija, A.; Van Dam, R.M.; Hu, F.B. Long-term coffee consumption and risk of cardiovascular disease: A systematic review and a dose-response meta-analysis of prospective cohort studies. *Circulation* **2014**, *129*, 643–659. [CrossRef]
21. Popiolek-Kalisz, J.; Fornal, E. Dietary Isorhamnetin Intake Is Inversely Associated with Coronary Artery Disease Occurrence in Polish Adults. *Int. J. Environ. Res. Public Health* **2022**, *19*, 12546. [CrossRef]
22. Williams, B.; Mancia, G.; Spiering, W.; Rosei, E.A.; Azizi, M.; Burnier, M.; Clement, D.L.; Coca, A.; De Simone, G.; Dominiczak, A.; et al. 2018 ESC/ESH Guidelines for the management of arterial hypertension. *Kardiol. Pol.* **2019**, *77*, 71–159. [CrossRef] [PubMed]
23. Dabeek, W.M.; Marra, M.V. Dietary quercetin and kaempferol: Bioavailability and potential cardiovascular-related bioactivity in humans. *Nutrients* **2019**, *11*, 2288. [CrossRef] [PubMed]
24. Xiao, J.; Muzashvili, T.S.; Georgiev, M.I. Advances in the biotechnological glycosylation of valuable flavonoids. *Biotechnol. Adv.* **2014**, *32*, 1145–1156. [CrossRef]
25. Galindo, P.; Rodriguez-Gómez, I.; González-Manzano, S.; Dueñas, M.; Jiménez, R.; Menéndez, C.; Vargas, F.; Tamargo, J.; Santos-Buelga, C.; Pérez-Vizcaíno, F.; et al. Glucuronidated quercetin lowers blood pressure in spontaneously hypertensive rats via deconjugation. *PLoS ONE* **2012**, *7*, e32673. [CrossRef]

26. Cogolludo, A.; Frazziano, G.; Briones, A.M.; Cobeño, L.; Moreno, L.; Lodi, F.; Salaices, M.; Tamargo, J.; Perez-Vizcaino, F. The dietary flavonoid quercetin activates BKCa currents in coronary arteries via production of H_2O_2. Role in vasodilatation. *Cardiovasc. Res.* **2007**, *73*, 424–431. [CrossRef]
27. Hussain, F.; Jahan, N.; Rahman, K.-U.; Sultana, B.; Jamil, S. Identification of hypotensive biofunctional compounds of Coriandrum sativum and evaluation of their Angiotensin-Converting Enzyme (ACE) inhibition potential. *Oxid. Med. Cell. Longev.* **2018**, *2018*, 4643736. [CrossRef] [PubMed]
28. Knekt, P.; Jarvinen, R.; Reunanen, A.; Maatela, J. Flavonoid intake and coronary mortality in Finland: A cohort study. *BMJ* **1996**, *312*, 478–481. [CrossRef]
29. Kondratiuk, V.E.; Synytsia, Y.P. Effect of quercetin on the echocardiographic parameters of left ventricular diastolic function in patients with gout and essential hypertension. *Wiad. Lek.* **2018**, *71*, 1554–1559.
30. Hertog, M.G.; Feskens, E.J.; Kromhout, D.; Hertog, M.G.; Hollman, P.C.; Hertog, M.G.; Katan, M. Dietary antioxidant flavonoids and risk of coronary heart disease: The Zutphen Elderly Study. *Lancet* **1993**, *342*, 1007–1011. [CrossRef]
31. Brüll, V.; Burak, C.; Stoffel-Wagner, B.; Wolffram, S.; Nickenig, G.; Müller, C.; Langguth, P.; Alteheld, B.; Fimmers, R.; Naaf, S.; et al. Effects of a quercetin-rich onion skin extract on 24 h ambulatory blood pressure and endothelial function in overweight-to-obese patients with (pre-)hypertension: A randomised double-blinded placebo-controlled cross-over trial. *Br. J. Nutr.* **2015**, *114*, 1263–1277. [CrossRef]
32. Kalus, U.; Pindur, G.; Jung, F.; Mayer, B.; Radtke, H.; Bachmann, K.; Mrowietz, C.; Koscielny, J.; Kiesewetter, H. Influence of the onion as an essential ingredient of the mediterranean diet on arterial blood pressure and blood fluidity. *Arzneimittelforschung* **2000**, *50*, 795–801. [CrossRef] [PubMed]
33. Mennen, L.I.; Sapinho, D.; De Bree, A.; Arnault, N.; Bertrais, S.; Galan, P.; Hercberg, S. Consumption of Foods Rich in Flavonoids Is Related to A Decreased Cardiovascular Risk in Apparently Healthy French Women. *J. Nutr.* **2004**, *134*, 923–926. [CrossRef] [PubMed]
34. Kawamoto, E.; Sakai, Y.; Okamura, Y.; Yamamoto, Y. Effects of boiling on the antihypertensive and antioxidant activities of onion. *J. Nutr. Sci. Vitaminol.* **2004**, *50*, 171–176. [CrossRef] [PubMed]

Systematic Review

Adherence to the DASH Diet and Risk of Hypertension: A Systematic Review and Meta-Analysis

Xenophon Theodoridis [1,2], Michail Chourdakis [1,*], Lydia Chrysoula [1], Violeta Chroni [1], Ilias Tirodimos [1], Konstantina Dipla [3], Eugenia Gkaliagkousi [2] and Areti Triantafyllou [2]

1 Laboratory of Hygiene, Social and Preventive Medicine and Medical Statistics, School of Medicine, Faculty of Health Sciences, Aristotle University of Thessaloniki, 54124 Thessaloniki, Greece; xtheodoridis@auth.gr (X.T.); lchrysoula@auth.gr (L.C.); violetac@auth.gr (V.C.); ityrodim@auth.gr (I.T.)
2 3rd Clinic of Internal Medicine, Papageorgiou Hospital, School of Medicine, Faculty of Health Sciences, Aristotle University of Thessaloniki, 56403 Thessaloniki, Greece; egkaliagkousi@auth.gr (E.G.); artriant@auth.gr (A.T.)
3 Exercise Physiology & Biochemistry Laboratory, Department of Sport Sciences at Serres, Aristotle University of Thessaloniki, 62110 Thessaloniki, Greece; kdipla@phed-sr.auth.gr
* Correspondence: mhourd@gapps.auth.gr; Tel.: +30-23-1099-9035

Abstract: The aim of this study was to assess the effect of the level of adherence to the DASH diet on hypertension risk by conducting a systematic review and meta-analysis. A systematic literature search was performed. Two independent investigators performed the study selection, data abstraction, and assessment of the included studies. The meta-analysis was performed separately with the adjusted hazard (HR) or incident rate ratios (IRR) and the odds ratios (OR) of the highest compared to the lowest DASH diet adherence scores using a random effects model. A total of 12 studies were included in the qualitative and quantitative synthesis. When cohort studies reporting HR were pooled together, high adherence to the DASH diet was associated with a lower risk of hypertension (HR: 0.81, 95% CI 0.73–0.90, I^2 = 69%, PI 0.61–1.08) compared to the low adherence. When cross-sectional studies reporting OR were combined, high adherence to the DASH diet was also related to a lower risk of hypertension (OR: 0.80, 95% CI 0.70–0.91, I^2 = 81%, PI 0.46–1.39). The findings suggest that high adherence to the DASH diet has a positive effect on reducing hypertension risk compared to low adherence. These data strengthen and are in line with all hypertension guidelines, indicating that lifestyle changes should start early even in populations with normal blood pressure.

Keywords: blood pressure; compliance; DASH diet; systematic review; meta-analysis; hypertension

Citation: Theodoridis, X.; Chourdakis, M.; Chrysoula, L.; Chroni, V.; Tirodimos, I.; Dipla, K.; Gkaliagkousi, E.; Triantafyllou, A. Adherence to the DASH Diet and Risk of Hypertension: A Systematic Review and Meta-Analysis. *Nutrients* **2023**, *15*, 3261. https://doi.org/10.3390/nu15143261

Academic Editor: Giuseppe Della Pepa

Received: 29 June 2023
Revised: 17 July 2023
Accepted: 18 July 2023
Published: 24 July 2023

Copyright: © 2023 by the authors. Licensee MDPI, Basel, Switzerland. This article is an open access article distributed under the terms and conditions of the Creative Commons Attribution (CC BY) license (https://creativecommons.org/licenses/by/4.0/).

1. Introduction

The prevalence of hypertension doubled in adults older than 30 from 1990 to 2019 [1]. According to the World Health Organization (WHO) [2], it is estimated that almost 50% of adults with hypertension are undiagnosed. Prediction models project that in 2030 approximately 40% of adults in the U.S.A. will develop some form of cardiovascular disease (CVD), including hypertension [3]. To tackle this public health issue, the WHO set a global target to ameliorate the prevalence of hypertension by $\frac{1}{4}$ by 2025 [4].

High blood pressure is the leading cause of disability-adjusted life years (DALY) [5] and accounts for most of the CVD events worldwide [6] and premature deaths [7]. Even though prevention policies are of fundamental importance to mitigate the tremendous increase in hypertension rates and reduce the development of related comorbidities, most clinical practice guidelines focused primarily on the treatment of this condition [8,9]. The recommendations for preventing hypertension include the amendment of dietary deviations from guidelines and physical inactivity [10]. Decreasing the amount of sodium intake and the achievement of weight loss for adults with overweight or obesity are strategies that will prevent hypertension-attributed deaths and reduce the risk of hypertension, respectively [6].

The most studied dietary pattern for high blood pressure is the Dietary Approaches to Stop Hypertension (DASH) eating plan. The DASH diet has been proposed for the management of high blood pressure due to the inclusion of food groups with antihypertensive properties [11]. More specifically, the DASH diet emphasizes the consumption of fruits, vegetables, whole grains, legumes, nuts, lean protein, and low-fat dairy products. Furthermore, it focuses on limited intakes of salt, added sugar, and saturated fat. Many studies have assessed whether the level of adherence to the DASH diet can reduce hypertension risk among the adult population, with inconclusive results [12–15].

To the best of our knowledge, there is not a published systematic review and meta-analysis assessing the effect of the level of adherence to the DASH diet on the development of hypertension. Hence, the objective of this work was to synthesize all the data from the available primary studies to shed light on the inconclusive results.

2. Materials and Methods

2.1. Protocol and Registration

The present systematic review and meta-analysis has been conducted according to the Preferred Reporting Items for Systematic Reviews and Meta-Analyses (PRISMA 2020) [16] and Meta-analysis Of Observational Studies in Epidemiology (MOOSE) [17] statements (Supplementary Tables S1 and S2). A pre-specified protocol has been registered in the Prospero repository (CRD42022344686).

2.2. Search Strategy

A comprehensive literature search was conducted in MEDLINE via PubMed, Scopus, and Web of Science Core Collection databases from inception to November 2022 by two independent reviewers. Furthermore, we inspected the references of the included studies for relevant articles. The grey literature was also searched for potential records. Finally, we consulted experts in the field of nutrition and hypertension for the provision of eligible studies. We used search terms such as "hypertension", "blood pressure", "DASH diet", and "dietary adherence". The full search string can be found in Supplementary Table S3.

2.3. Study Selection

We included records that met the following criteria: (1) observational and/or interventional studies, (2) including adult population without a hypertension diagnosis, (3) comparing the effect of high and low adherence to the DASH diet, (4) on the risk of developing hypertension.

Adherence to the DASH diet is defined as the degree to which an individual follows the DASH diet. We defined hypertension according to ESC/ESH guidelines [8], namely, systolic blood pressure (SBP) $\geq$ 140 or/and diastolic blood pressure (DBP) $\geq$ 90 or the use of antihypertensive medication.

Studies including pregnant or pediatric populations or those written in a non-English language were excluded.

2.4. Data Extraction

We independently abstracted data regarding study characteristics including the first author's name, publication year, the country in which the study took place, study design, study population details, comorbidities, DASH diet assessment tool, and the use of anti-hypertensive medication. As far as statistical data are concerned, we independently extracted risk estimates with their corresponding 95% confidence intervals (CI) regarding the risk of hypertension based on the level (high or low) of adherence to the DASH diet.

2.5. Risk of Bias in Individual Studies

Two independent researchers assessed the risk of bias in the included observational studies using the Newcastle–Ottawa scale (NOS) and checklists for cross-sectional or cohort studies developed by the Joanna Briggs Institute (JBI). The JBI checklist for the cohort studies includes 11 items regarding the study design, while the checklist for the cross-sectional studies comprises 8 questions. There are three available options to respond to these items, "yes" indicating high quality, "no" indicating poor quality, or "unclear".

2.6. Data Synthesis

To answer our research question, we conducted two statistical analyses, one including only cohort studies and another one including only cross-sectional studies. The meta-analysis was performed separately for the adjusted hazard ratios (HR) or incident rate ratios (IRR) and odds ratios (OR) of the highest compared to the lowest DASH diet adherence score using a random effects model. The heterogeneity was estimated using the estimator proposed by Paule and Mandel [18], and measured using the I^2 index, which describes the percentage of variability due to heterogeneity rather than sampling error, and the τ^2 [19]. We present the prediction interval (PI), which facilitates clinical interpretation of the heterogeneity and quantifies the range of the effect size that a future study will fall [20]. Funnel plots and publication bias tests for assessing their asymmetry were not calculated due to the few included studies [21]. We also performed a subgroup and sensitivity analysis in order to explain heterogeneity and assess the robustness of our findings, respectively. Data were analyzed using the R Studio software (version 2023.06.0) and meta package. Statistical significance was set at $p < 0.05$.

2.7. Quality of the Evidence

The certainty of the evidence of our findings was assessed using the Grading of Recommendations, Assessment, Development, and Evaluations (GRADE) approach, as recommended by the Cochrane Handbook [19].

3. Results

3.1. Database Search and Study Characteristics

An electronic literature search was performed on MEDLINE via PubMed, Scopus, and Web of Science Core Collection, and a total of 4319 articles were retrieved. After the duplicate removal process, 628 records remained for further evaluation. After title and abstract screening, 136 articles were assessed for eligibility. The final sample of the study incorporated 12 individual studies [14,15,22–31]. Figure 1 presents the details of the study search and selection process.

3.2. Definitions of the DASH Diet

Four studies used the definition of the DASH diet based on the DASH score constructed by Fung and colleagues [32], while six studies provided their own definition of the DASH diet score based either on food groups or types of macro- and micronutrients using different cut points for low and high adherence. One study defined the DASH diet according to recommended and restricted food groups as well as sodium consumption based on the guide published by the National Institutes of Health and the National Heart Lung and Blood Institute. In one study, the DASH diet was described as the sum of three food groups, namely, vegetables, fruit, and milk products using the hypothesis-oriented pattern variable.

3.3. Outcome of Interest

When the cohort studies reporting HR were pooled together (Figure 2), high adherence to the DASH diet was associated with a lower risk of hypertension (HR: 0.81, 95% CI 0.73–0.90, $I^2 = 69\%$, PI 0.61–1.08) compared to low adherence. Based on the Cochrane Handbook, the heterogeneity appears to be substantial.

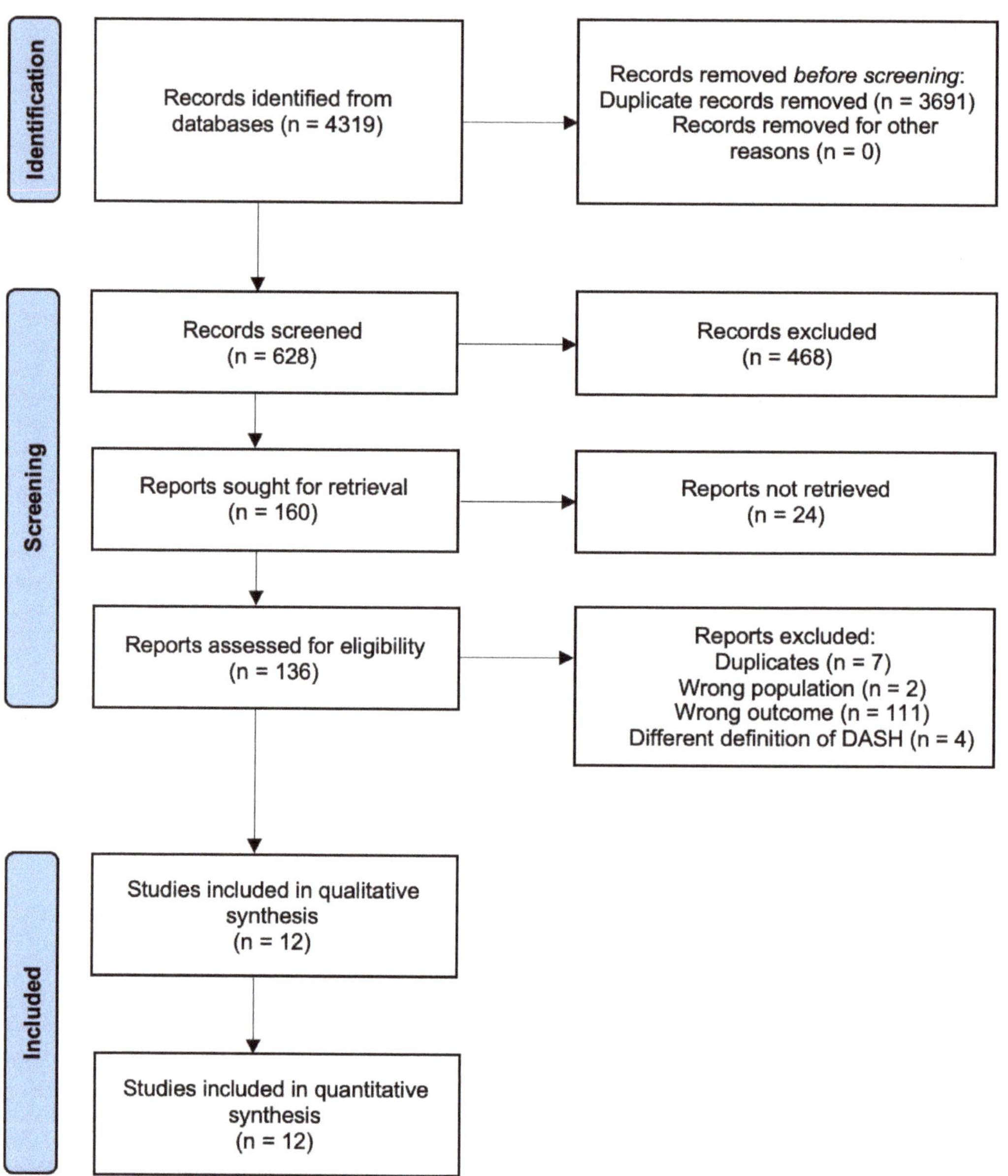

Figure 1. Flow diagram of the eligibility process.

The main characteristics of the included studies are summarized in Tables 1 and 2. Briefly, of the total 12 studies, three were cross-sectional and nine were cohort studies. The total number of participants exceeded 115,000 and their mean age ranged from 36.3 to 61.0 years old.

Table 1. Characteristics of the included studies.

Study ID, Country	Study Design, Effect Size	Population	No of Participants (Low/High Adherence)	Mean Age ± SD	Exclusion Criteria	DASH Assessment Tool	Hypertension Diagnosis
Bai et al., 2017 [22], China	Longitudinal-cohort, HR	Chinese adults	-	42 ± 9.3	Younger than 18 years old, missing average SBP or DBP, identified hypertension, antihypertensive medication, existing diagnosis of diabetes, myocardial infarction, or apoplexy	DASH diet score Fung et al. (2008) [32]	SBP ≥ 140 or DBP ≥ 90 or antihypertensive medication use
Camões et al., 2010 [23], Portugal	Longitudinal cohort study, HR	Portuguese adults, resident in Porto and at risk of developing hypertension	246/256	-	Age < 39 years old, missing information on BP measurements, hypertensive	Developed DASH diet score	SBP ≥ 140 or DBP ≥ 90 or antihypertensive medication use
Cherfan et al., 2019 [30], France	Cross-sectional analysis, OR	Adult workers or retired	3709/29,342	-	BMI < 18 kg/m^2	Constructed DASH diet score according to Fung et al. (2008) [32]	SBP ≥ 140 or DBP ≥ 90 or antihypertensive medication use
Folsom et al., 2007 [24], U.S.	Cohort study, HR	Adult women	5017/4041	61.0	History of physician-diagnosed hypertension, heart attack, angina, heart disease, diabetes, more than 29 blank items on FFQ, EI < 500 kcal or >5000 kcal, missing covariates	Developed DASH diet index score	Self-report of hypertension
Francisco et al., 2020 [25], Brazil	Longitudinal cohort study, HR	Adults active or retired civil services of higher research institutions located in Brazil	4987/645	49.9 ± 8.3	Hypertension, antihypertensive drugs, CVD, changed dietary habits in the last 6 months, self-reported their race/skin color as Asian, Brazilian indigenous, missing information on BP values	Dash diet score calculated by National Institutes of Health, National Heart Lung and Blood Institute	SBP ≥ 140 or DBP ≥ 90 or antihypertensive medication use
Jiang et al., 2015 [14], U.S.	Longitudinal cohort study, HR	U.S. adults	585/331	52.5 ± 9.4	Medication, no SBP or DPB recorded at follow-ups, no valid FFQ, missing information for hypertension medication	Developed DASH diet score using score by Folsom et al. (2007) [24]	SBP ≥ 140 or DBP ≥ 90 or antihypertensive medication use
Kanauchi et al., 2015 [15], Japan	Cross-sectional, OR	Male workers	-	45.3 ± 6.9	Diabetes, CKD, implausibly low or high estimated EI, missing information	Developed DASH diet score	SBP ≥ 140 or DBP ≥ 90
Lelong et al., 2017 [26], France	Prospective cohort study, HR	Adults internet user volunteer	19,967/ 19,323	41.9 ± 14.0	Energy under reporters, with < 3 24 h valid recalls, prevalent hypertension, cancer, diabetes mellitus, and cardiovascular disease, pregnant women, missing or invalid data on health status, anthropometric measurements, or physical activity	DASH diet score Fung et al. (2008) [32]	Self-report of hypertension
Li et al., 2016 [27], U.S.	Cohort study, HR	Adult women	706/747	36.5 ± 4.3	History of cardiovascular disease, cancer, multiple gestations or pregnancies lasting <6 months, history of GDM, history of hypertension beforethe diagnosis of GDM or with missing data on post-pregnancy diet	DASH diet score Fung et al. (2008) [32]	Self-report of hypertension

Table 1. *Cont.*

Study ID, Country	Study Design, Effect Size	Population	No of Participants (Low/High Adherence)	Mean Age ± SD	Exclusion Criteria	DASH Assessment Tool	Hypertension Diagnosis
Schulze et al., 2003 [28], Germany	Cohort study, HR	Women in the EPIC-Potsdam Study	-	-	Previous diagnosis of hypertension, antihypertensive medication within a 4-week period prior to the baseline examination, missing information on dietary intake, estimated basal metabolic rate, physical activity, lifestyle characteristics, and anthropometric measurements; current pregnancy or breastfeeding, outlying total energy intake, with no follow-up, possible hypertension for whom we did not have completed verification, prevalent or secondary hypertension	DASH diet score based on hypothesis-oriented pattern variable	-
Toledo et al., 2010 [29], Spain	Prospective cohort study, HR	University graduates	6487/158	36.3 ± 11.0	Self-reported prevalent hypertension with extreme total EI, prevalent CVD at baseline	Developed DASH diet score	Self-report of hypertension
Yang et al., 2022 [31], China	Cross-sectional, OR	Chinese adults	12,298/11,862	-	Incomplete dietary information, incomplete basic information, incomplete physical examination and laboratory test, implausible dietary EI < 500 kcal/d or >5000 kcal/d, and pre-diagnosed coronary heart disease or stroke	Developed DASH diet score	SBP $\geq$ 140 or DBP $\geq$ 90 or antihypertensive medication use

BMI: Body Mass Index; BP: Blood Pressure; CKD: Chronic Kidney Disease; CVD: Cardiovascular Disease; DASH: Dietary Approaches to Stop Hypertension; DBP: Diastolic Blood Pressure; EI: Energy Intake; EPIC: European Prospective Investigation into Cancer and Nutrition; ESC/ESH: European Society of Cardiology/European Society of Hypertension; FFQ: Food Frequency Questionnaire; GDM: Gestational Diabetes Mellitus; HR: Hazard Ratio; OR: Odds Ratio; SBP: Systolic Blood Pressure; SD: Standard Deviation.

Table 2. Patients' health characteristics of the included studies.

First Author, Year	BMI	SBP	DBP	Physical Activity	Smoking Status	Sodium Intake	Potassium Intake
	(Low/High)	(Low/High)	(Low/High)	(Low/High)	(Low/High)	(Low/High)	(Low/High)
Bai et al., 2017 [22]	NA	NA	NA	NA	NA	NA	NA
Camões et al., 2010 [23]	NA	NA	NA	NA	NA	NA	NA
Cherfan et al., 2019 [30]	NA	NA	NA	NA	NA	NA	NA
Folsom et al., 2007 [24]	26.3/25.3	NA	NA	16.0%/40.0% high PA *	Current smokers = 22.0%/10.0%	2124.0 mg/d 2275.0 mg/d	1147.0 mg/d 1437.0 mg/d
Francisco et al., 2020 [25]	25.8 ± 4.2/ 24.9 ± 3.8	114.5 ± 11.5/ 114.5 ± 11.8	72.7 ± 8.1/ 71.4 ± 8.2	Low Adherence: Light = 78.6% Moderate = 14.1% High = 7.3% High Adherence: Light = 62.8% Moderate = 24.8% High = 12.4%	Low Adherence: Non-smoker = 58.8% Former smoker = 25.8% Smoker = 15.4% High Adherence: Non-smoker = 65.3% Former smoker = 25.4% Smoker = 9.3%	4.6 ± 14.4 g/d 3.5 ± 3.0 g/d	3982.0 ± 1607.0 mg/d 5260.0 ± 1664.0 mg/d

Table 2. *Cont.*

First Author, Year	BMI	SBP	DBP	Physical Activity	Smoking Status	Sodium Intake	Potassium Intake
	(Low/High)	(Low/High)	(Low/High)	(Low/High)	(Low/High)	(Low/High)	(Low/High)
Jiang et al., 2015 [14]	27.1/25.9	121.1/119.0	73.6/71.7	35.6/34.6 PAI	35.9%/7.0%	1145.3/1000 kcal 1146.0/1000 kcal	1468.3/1000 kcal 1902.2/1000 kcal
Kanauchi et al., 2015 [15]	NA	NA	NA	NA	NA	NA	NA
Lelong et al., 2017 [26]	23.8 ± 4.7/ 22.7 ± 3.6	NA	NA	Low Adherence: Low = 31.3% Moderate = 41.5% High = 27.3% High Adherence: Low = 17.4% Moderate = 44.1% High = 38.5%	Low Adherence: Never = 48.7% Former Smoker = 25.8% Current = 25.6% High Adherence: Never = 53.6% Former Smoker = 36.1% Current = 38.5%	2907.0 ± 958.0 mg/d 2454.0 ± 857.0 mg/d	2623.0 ± 726.0 mg/d 3409.0 ± 884.0 mg/d
Li et al., 2016 [27]	26.8 ± 6.5/ 25.8 ± 5.7	NA	NA	12.5 ± 18.3/21.9 ± 25.4 (MET × h/week)	19.0%/7.0%	NA	NA
Schulze et al., 2003 [28]	NA	NA	NA	NA	NA	NA	NA
Toledo et al., 2010 [29]	23.0 ± 3.0/ 23.0 ± 3.0	NA	NA	23.5 ± 20.9/32.1 ± 30.1 (MET × h/week)	Low Adherence: Current = 25.0% Ex-smokers = 25.0% High Adherence: Current = 15.0% Ex-smokers = 30.0%	3.4 ± 2.2 g/d 3.1 ± 1.5 g/d	4.3 ± 1.2 g/d 7.3 ± 2.1 g/d
Yang et al., 2022 [31]	NA	NA	NA	NA	NA	NA	NA

MET: Metabolic Equivalent of Task; NA: Not Available; PA: Physical Activity; PAI: Physical Activity Index. * No definition of high PA.

Study	logHR	SE(logHR)	HR	95%-CI	Weight (common)	Weight (random)
Bai et al., 2017	-0.1625	0.0751	0.85	[0.73; 0.98]	17.2%	16.8%
Camões et al., 2010	-0.1744	0.2115	0.84	[0.55; 1.26]	2.2%	5.2%
Folsom et al., 2007	-0.0305	0.0528	0.97	[0.87; 1.07]	34.9%	20.1%
Francisco et al., 2020	-0.2107	0.1279	0.81	[0.63; 1.04]	5.9%	10.5%
Jiang et al., 2015	-0.0726	0.1137	0.93	[0.73; 1.14]	7.5%	11.9%
Lelong et al., 2017	-0.4155	0.0656	0.66	[0.58; 0.75]	22.6%	18.2%
Li et al., 2016	-0.3285	0.1121	0.72	[0.58; 0.90]	7.7%	12.1%
Schulze et al., 2003	-0.3567	0.2657	0.70	[0.42; 1.19]	1.4%	3.6%
Toledo et al., 2010	-0.7340	0.4201	0.48	[0.21; 1.09]	0.6%	1.6%
Common effect model			**0.83**	**[0.78; 0.88]**	**100.0%**	--
Random effects model			**0.81**	**[0.73; 0.90]**	--	**100.0%**
Prediction interval				**[0.61; 1.08]**		

Heterogeneity: $I^2 = 69\%$, $\tau^2 = 0.0119$, $p < 0.01$ — Hazard Ratio axis: 0.5, 1, 2 — Favors High Adherence / Favors Low Adherence

Figure 2. Forest plot for the hypertension risk when cohort studies were pooled together. Bai et al., 2017 [22], Camões et al., 2010 [23], Folsom et al., 2007 [24], Francisco et al., 2020 [25], Jiang et al., 2015 [14], Lelong et al., 2017 [26], Li et al., 2016 [27], Schulze et al., 2003 [28], Toledo et al., 2010 [29].

When cross-sectional studies reporting OR were combined (Figure 3), high adherence to the DASH diet was not related to the risk of hypertension (OR: 0.80, 95% CI 0.70–0.91, $I^2 = 81\%$, PI 0.46–1.39). A considerable heterogeneity was observed for the DBP outcome.

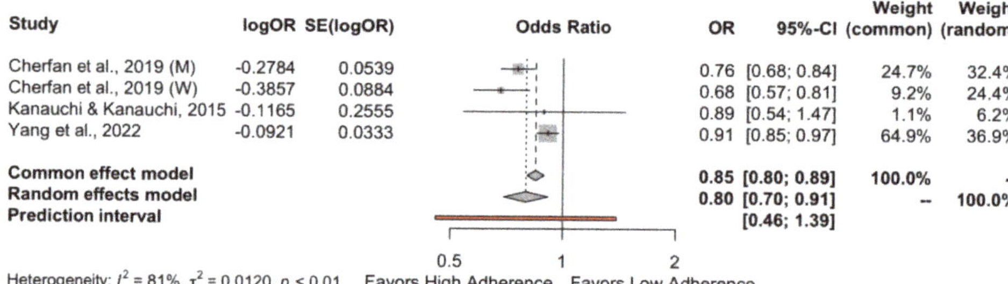

Figure 3. Forest plot for the hypertension risk when cross-sectional studies were pooled together. Cherfan et al., 2019 (M) [30], Cherfan et al., 2019 (W) [30], Kanauchi & Kanauchi, 2015 [15], Yang et al., 2022 [31].

3.4. Risk of Bias Assessment

The results of the assessment using the NOS are presented in Supplementary Tables S4 and S5. The quality assessment of the cross-sectional and cohort studies based on the JBI checklists are presented in Supplementary Tables S6 and S7, respectively. For the cross-sectional studies, the articles did not provide enough information about the study subjects and the setting. Regarding the appraisal of the cohort studies, only two studies [22,27] had sufficient follow-up time for the outcome of interest to occur.

3.5. Subgroup and Sensitivity Analysis

We conducted a subgroup analysis based on the hypertension diagnosis (i.e., SBP $\geq$ 140 and/or DBP $\geq$ 90 or antihypertensive medication use versus self-report of hypertension). The results of the subgroup analysis indicate that there is no difference between the two methods used for hypertension diagnosis. Specifically, the pooled estimate for diagnosis of hypertension based on values of SBP $\geq$ 140 and/or DBP $\geq$ 90 or the use of antihypertensive medication was HR: 0.85 (95% CI 0.77–0.95, $I^2 = 0\%$). On the other hand, the summary effect for the self-report of hypertension method was HR: 0.75 (95% CI 0.60–0.95, $I^2 = 88\%$) (Supplementary Figure S1).

A sensitivity analysis has also been conducted to assess the robustness of the findings. In this analysis, cohort studies with a NOS score < 7 were removed. The results of the sensitivity analysis were HR: 0.81, 95% CI 0.71–0.92, I^2 = 4%, PI 0.64–1.01 (Supplementary Figure S2).

3.6. Certainty of the Evidence

According to the GRADE approach, the quality of our evidence was deemed very low for both effect sizes (i.e., HR and OR). Risk of bias, inconsistency, indirectness, and imprecision were the domains that both comparisons were downgraded by one level.

4. Discussion

The present systematic review and meta-analysis aimed to evaluate the effect of the level of adherence to the DASH diet on hypertension risk. The findings suggest that, based on the pooled estimate from the cohort studies, high adherence to the DASH diet has a positive effect on hypertension prevention compared to low adherence. This observation is in line with the findings resulting from the data of the cross-sectional studies that were also synthesized.

With respect to potential antihypertensive mechanisms of the DASH diet, decreased sodium and increased potassium intake are among the most well-studied factors. Specifically, the DASH diet is rich in fruits and vegetables with high amounts of potassium, which shows vasoactive properties and possibly reduces blood pressure through a decrease in vascular smooth muscle contraction [33]. On the other hand, potassium increases urinary sodium excretion and reduces insulin resistance and oxidative damage [25]. Insulin resistance with compensatory hyperinsulinemia and reactive oxygen species that influence the homeostasis of the vascular wall could lead to hypertension [34,35].

On the contrary, high sodium diets lead to water retention, which, in turn, causes an expansion in circulating volumes, a rise in cardiac output, and an increase in kidney perfusion pressure [36]. Moreover, high kidney perfusion pressure prompts a rise in the glomerular filtration rate and sodium excretion in order to restore body fluids. Another plausible mechanism is that excessive sodium intake elicits a reduction in vascular nitric oxide concentration, which is responsible for endothelium-dependent dilation [37].

High dietary sodium intake is associated with arterial stiffness mainly due to a modification in the extracellular matrix of the arterial wall [38,39]. A J-shaped curve has been found to resemble the relationship between sodium or potassium intake and vascular structure and function [40]. Evidence supports that arterial stiffness is related to a higher risk of hypertension incidence [41].

An increase in dietary fiber intake has also been associated with a reduction in both systolic and diastolic blood pressure [42]. The reduction in blood pressure depends on the type of dietary fiber, where β-glucan appears to be the most effective one [43]. An improvement of insulin sensitivity and endothelial function, stimulation of the absorption of minerals in the gastrointestinal tract, and reduction in body weight are among the mechanisms that have been proposed to link fiber intake and blood pressure control [44].

A systematic review and meta-analysis of randomized controlled trials demonstrated that the DASH diet reduces blood pressure in both normotensive and hypertensive adults [11]. This study also showed that the blood pressure-lowering effect of the DASH diet was more prominent in participants aged <50 years and among those with a sodium intake >2400 mg/d [11]. Another recently published systematic review and meta-analysis of randomized controlled trials found that a modified DASH diet is effective in decreasing blood pressure and some cardiometabolic markers, such as waist circumference and triglyceride concentration in patients with hypertension [45]. From this study, a higher baseline blood pressure is linked to more pronounced systolic and diastolic blood pressure decreases [45]. Finally, another systematic review and dose-response meta-analysis by Soltani and colleagues [46] indicated that even a low adherence to the DASH diet was associated with lower all-cause, cardiovascular, and cancer mortality.

Our findings showed that high adherence to the DASH diet has a protective role on the risk of hypertension in comparison with low adherence. Even though the pooled estimates from the cohort and cross-sectional studies are in agreement, findings derived from the cross-sectional studies should be interpreted with more caution, as they are at a lower level of the evidence hierarchy compared to the cohort studies [47]. Hence, these studies may have less methodological rigor and more biases affecting their conclusions. This is also supported by the wider PI emerging from the synthesis of the cross-sectional studies when compared to the PI resulting from the pooling of the cohort studies [48].

To further explore the substantial heterogeneity presented in the synthesis of the cohort studies, a subgroup analysis based on the hypertension diagnostic method was performed. The results of this analysis showed that there was no statistical heterogeneity between studies that used the most accurate diagnostic method for hypertension. Contrarily, high heterogeneity was still present in the studies that used self-reporting of hypertension as the method of their choice.

The results of the sensitivity analysis are in line with the results of our primary analysis, indicating that our findings are robust. Furthermore, upon exclusion of the cohort studies deemed of low quality based on the NOS assessment, a reduction in the heterogeneity of the summary effect to an I^2 = 4% was observed. This reduction indicates the absence of heterogeneity among the included studies.

The findings of the present systematic review indicate the beneficial effect of high adherence to the DASH diet on the risk of developing hypertension in subjects with normal blood pressure values. Healthcare professionals including doctors, dietitians, and nurses, as well as policy-makers, should recommend early compliance to the basic guidelines of the DASH diet in order to reduce the incidence of hypertension and the related comorbidities. Future studies should prioritize the development and validation of an instrument assessing adherence to the DASH diet, which could be utilized in research trials. Upon such a successful acceptance from the scientific society, it could then be also applied to the clinical setting. Additionally, larger sample sizes studies encompassing diverse participants are welcomed.

Compliance with the Cochrane guidelines, the rigor of statistical and methodological aspects used, and that this is the first systematic review and meta-analysis assessing the effect of the level of adherence to the DASH diet on hypertension risk in normotensive individuals are some of the strengths of our study. However, there are limitations that should be accounted for. Firstly, the low quality of the included observational studies reduces the certainty of the evidence. Furthermore, some studies reported hazard ratios while others reported odds ratios; hence, we could not pool data from all the available studies. Another limitation is that the included studies defined hypertension and DASH diet adherence based on different thresholds and scores, respectively. Lastly, the inclusion of studies written in the English language can only comprise a limitation of our study. However, two meta-epidemiologic studies showed that restricting evidence synthesis to English-language articles has a modest effect on effect estimates and the study's conclusion [49,50].

5. Conclusions

The findings suggest that high adherence to the DASH diet has a positive effect on reducing hypertension risk compared to low adherence. These data strengthen and are totally in line with all hypertension guidelines, i.e., European, American, and International, independent of the cut-off points used to define hypertension, pointing out that lifestyle modifications should start early before the establishment of hypertension, even in subjects with normal blood pressure levels.

Supplementary Materials: The following supporting information can be downloaded at: https://www.mdpi.com/article/10.3390/nu15143261/s1, Supplementary Table S1. PRISMA 2020 checklist. Supplementary Table S2. MOOSE checklist. Supplementary Table S3. Search strategy for identifying studies on Pubmed. Supplementary Table S4. Quality assessment using the Newcastle-Ottawa scale for cross-sectional studies. Supplementary Table S5. Quality assessment using the Newcastle-Ottawa scale for cohort studies. Supplementary Table S6. Quality assessment using the JBI checklist for cross-sectional studies. Supplementary Table S7. Quality assessment using the JBI checklist for cohort studies. Supplementary Figure S1. Subgroup analysis for the cohort studies based on hypertension diagnosis. Supplementary Figure S2. Sensitivity analysis by removing cohort studies with a NOS score < 7.

Author Contributions: Conceptualization, X.T. and A.T.; methodology, X.T.; formal analysis, X.T.; investigation, L.C., I.T. and V.C.; data curation, X.T. and L.C.; writing—original draft preparation, X.T. and L.C.; writing—review and editing, M.C., K.D., E.G. and A.T.; visualization, X.T.; supervision, M.C. and A.T. All authors have read and agreed to the published version of the manuscript.

Funding: This research received no external funding.

Institutional Review Board Statement: Not applicable.

Informed Consent Statement: Not applicable.

Data Availability Statement: The data presented in this study are available in the main text and Supplementary Materials.

Conflicts of Interest: The authors declare no conflict of interest.

References

1. Zhou, B.; Carrillo-Larco, R.M.; Danaei, G.; Riley, L.M.; Paciorek, C.J.; Stevens, G.A.; Gregg, E.W.; Bennett, J.E.; Solomon, B.; Singleton, R.K.; et al. Worldwide Trends in Hypertension Prevalence and Progress in Treatment and Control from 1990 to 2019: A Pooled Analysis of 1201 Population-Representative Studies with 104 Million Participants. *Lancet* **2021**, *398*, 957–980. [CrossRef]
2. World Health Organization. Hypertension. Available online: https://www.who.int/news-room/fact-sheets/detail/hypertension (accessed on 27 November 2022).
3. Heidenreich, P.A.; Trogdon, J.G.; Khavjou, O.A.; Butler, J.; Dracup, K.; Ezekowitz, M.D.; Finkelstein, E.A.; Hong, Y.; Johnston, S.C.; Khera, A.; et al. Forecasting the Future of Cardiovascular Disease in the United States. *Circulation* **2011**, *123*, 933–944. [CrossRef] [PubMed]
4. Kulkarni, S. Hypertension Management in 2030: A Kaleidoscopic View. *J. Hum. Hypertens.* **2020**, *35*, 812–817. [CrossRef] [PubMed]
5. Mensah, G.A. Epidemiology and Global Burden of Hypertension. In *The ESC Textbook of Cardiovascular Medicine*; Came, J., Lüscher, T., Maurer, G., Serruys, P., Eds.; Oxford University Press: Oxford, UK, 2018; pp. 290–297.
6. Carey, R.M.; Muntner, P.; Bosworth, H.B.; Whelton, P.K. Prevention and Control of Hypertension: JACC Health Promotion Series. *J. Am. Coll. Cardiol.* **2018**, *72*, 1278. [CrossRef] [PubMed]
7. Campbell, N.R.C.; Whelton, P.K.; Orias, M.; Wainford, R.D.; Cappuccio, F.P.; Ide, N.; Neal, B.; Cohn, J.; Cobb, L.K.; Webster, J.; et al. 2022 World Hypertension League, Resolve to Save Lives and International Society of Hypertension Dietary Sodium (Salt) Global Call to Action. *J. Hum. Hypertens.* **2022**, *37*, 428–437. [CrossRef] [PubMed]
8. Williams, B.; Mancia, G.; Spiering, W.; Agabiti Rosei, E.; Azizi, M.; Burnier, M.; Clement, D.L.; Coca, A.; de Simone, G.; Dominiczak, A.; et al. 2018 ESC/ESH Guidelines for the Management of Arterial HypertensionThe Task Force for the Management of Arterial Hypertension of the European Society of Cardiology (ESC) and the European Society of Hypertension (ESH). *Eur. Heart J.* **2018**, *39*, 3021–3104. [CrossRef]
9. Unger, T.; Borghi, C.; Charchar, F.; Khan, N.A.; Poulter, N.R.; Prabhakaran, D.; Ramirez, A.; Schlaich, M.; Stergiou, G.S.; Tomaszewski, M.; et al. 2020 International Society of Hypertension Global Hypertension Practice Guidelines. *Hypertension* **2020**, *75*, 1334–1357. [CrossRef]
10. Whelton, P.K.; Carey, R.M.; Aronow, W.S.; Casey, D.E.; Collins, K.J.; Himmelfarb, C.D.; DePalma, S.M.; Gidding, S.; Jamerson, K.A.; Jones, D.W.; et al. 2017 ACC/AHA/AAPA/ABC/ACPM/AGS/APhA/ ASH/ASPC/NMA/PCNA Guideline for the Prevention, Detection, Evaluation, and Management of High Blood Pressure in Adults a Report of the American College of Cardiology/American Heart Association Task Force on Clinical Practice Guidelines. *Hypertension* **2018**, *71*, E13–E115. [CrossRef]
11. Filippou, C.D.; Tsioufis, C.P.; Thomopoulos, C.G.; Mihas, C.C.; Dimitriadis, K.S.; Sotiropoulou, L.I.; Chrysochoou, C.A.; Nihoyannopoulos, P.I.; Tousoulis, D.M. Dietary Approaches to Stop Hypertension (DASH) Diet and Blood Pressure Reduction in Adults with and without Hypertension: A Systematic Review and Meta-Analysis of Randomized Controlled Trials. *Adv. Nutr.* **2020**, *11*, 1150. [CrossRef]
12. Günther, A.L.B.; Liese, A.D.; Bell, R.A.; Dabelea, D.; Lawrence, J.M.; Rodriguez, B.L.; Standiford, D.A.; Mayer-Davis, E.J. Association between the Dietary Approaches to Hypertension Diet and Hypertension in Youth with Diabetes Mellitus. *Hypertension* **2009**, *53*, 6–12. [CrossRef]

13. Hassani Zadeh, S.; Salehi-Abargouei, A.; Mirzaei, M.; Nadjarzadeh, A.; Hosseinzadeh, M. The Association between Dietary Approaches to Stop Hypertension Diet and Mediterranean Diet with Metabolic Syndrome in a Large Sample of Iranian Adults: YaHS and TAMYZ Studies. *Food Sci. Nutr.* **2021**, *9*, 3932. [CrossRef]
14. Jiang, J.; Liu, M.; Troy, L.M.; Bangalore, S.; Hayes, R.B.; Parekh, N. Concordance with DASH Diet and Blood Pressure Change: Results from the Framingham Offspring Study (1991–2008). *J. Hypertens.* **2015**, *33*, 2223–2230. [CrossRef]
15. Kanauchi, M.; Kanauchi, K. Diet Quality and Adherence to a Healthy Diet in Japanese Male Workers with Untreated Hypertension. *BMJ Open* **2015**, *5*, e008404. [CrossRef] [PubMed]
16. Page, M.J.; McKenzie, J.E.; Bossuyt, P.M.; Boutron, I.; Hoffmann, T.C.; Mulrow, C.D.; Shamseer, L.; Tetzlaff, J.M.; Akl, E.A.; Brennan, S.E.; et al. The PRISMA 2020 Statement: An Updated Guideline for Reporting Systematic Reviews. *BMJ* **2021**, *372*, n71. [CrossRef]
17. Stroup, D.F.; Berlin, J.A.; Morton, S.C.; Olkin, I.; Williamson, G.D.; Rennie, D.; Moher, D.; Becker, B.J.; Sipe, T.A.; Thacker, S.B. Meta-Analysis of Observational Studies in Epidemiology: A Proposal for Reporting. Meta-Analysis of Observational Studies in Epidemiology (MOOSE) Group. *JAMA* **2000**, *283*, 2008–2012. [CrossRef] [PubMed]
18. Veroniki, A.A.; Jackson, D.; Viechtbauer, W.; Bender, R.; Bowden, J.; Knapp, G.; Kuss, O.; Higgins, J.P.; Langan, D.; Salanti, G. Methods to Estimate the Between-Study Variance and Its Uncertainty in Meta-Analysis. *Res. Synth. Methods* **2016**, *7*, 55–79. [CrossRef] [PubMed]
19. Higgins, J.P.T.; Thomas, J.; Chandler, J.; Cumpston, M.; Li, T.; Page, M.J.; Welch, V.A. *Cochrane Handbook for Systematic Reviews of Interventions*; Wiley: Hoboken, NJ, USA, 2019; ISBN 9781119536604.
20. IntHout, J.; Ioannidis, J.P.A.; Rovers, M.M.; Goeman, J.J. Plea for Routinely Presenting Prediction Intervals in Meta-Analysis. *BMJ Open* **2016**, *6*, e010247. [CrossRef] [PubMed]
21. Page, M.J.; Sterne, J.A.C.; Higgins, J.P.T.; Egger, M. Investigating and Dealing with Publication Bias and Other Reporting Biases in Meta-Analyses of Health Research: A Review. *Res. Synth. Methods* **2021**, *12*, 248–259. [CrossRef] [PubMed]
22. Bai, G.; Zhang, J.; Zhao, C.; Wang, Y.; Qi, Y.; Zhang, B. Adherence to a Healthy Lifestyle and a DASH-Style Diet and Risk of Hypertension in Chinese Individuals. *Hypertens. Res.* **2017**, *40*, 196–202. [CrossRef]
23. Camões, M.; Oliveira, A.; Pereira, M.; Severo, M.; Lopes, C. Role of Physical Activity and Diet in Incidence of Hypertension: A Population-Based Study in Portuguese Adults. *Eur. J. Clin. Nutr.* **2010**, *64*, 1441–1449. [CrossRef]
24. Folsom, A.R.; Parker, E.D.; Harnack, L.J. Degree of Concordance with DASH Diet Guidelines and Incidence of Hypertension and Fatal Cardiovascular Disease. *Am. J. Hypertens.* **2007**, *20*, 225–232. [CrossRef] [PubMed]
25. Francisco, S.C.; Araújo, L.F.; Griep, R.H.; Chor, D.; Molina, M.D.C.B.; Mil, J.G.; Bensenor, I.M.; Matos, S.M.A.; Barreto, S.M.; Giatti, L. Adherence to the Dietary Approaches to Stop Hypertension (DASH) and Hypertension Risk: Results of the Longitudinal Study of Adult Health (ELSA-Brasil). *Br. J. Nutr.* **2020**, *123*, 1068–1077. [CrossRef] [PubMed]
26. Lelong, H.; Blacher, J.; Baudry, J.; Adriouch, S.; Galan, P.; Fezeu, L.; Hercberg, S.; Kesse-Guyot, E. Individual and Combined Effects of Dietary Factors on Risk of Incident Hypertension Prospective Analysis from the Nutrinet-Santé Cohort. *Hypertension* **2017**, *70*, 712–720. [CrossRef]
27. Li, S.; Zhu, Y.; Chavarro, J.E.; Bao, W.; Tobias, D.K.; Ley, S.H.; Forman, J.P.; Liu, A.; Mills, J.; Bowers, K.; et al. Healthful Dietary Patterns and the Risk of Hypertension among Women with a History of Gestational Diabetes Mellitus: A Prospective Cohort Study. *Hypertension* **2016**, *67*, 1157–1165. [CrossRef] [PubMed]
28. Schulze, M.B.; Hoffmann, K.; Kroke, A.; Boeing, H. Risk of Hypertension among Women in the EPIC-Potsdam Study: Comparison of Relative Risk Estimates for Exploratory and Hypothesis-Oriented Dietary Patterns. *Am. J. Epidemiol.* **2003**, *158*, 365–373. [CrossRef]
29. Toledo, E.; Carmona-Torre, F.D.A.; Alonso, A.; Puchau, B.; Zulet, M.A.; Martinez, J.A.; Martinez-Gonzalez, M.A. Hypothesis-Oriented Food Patterns and Incidence of Hypertension: 6-Year Follow-up of the SUN (Seguimiento Universidad de Navarra) Prospective Cohort. *Public Health Nutr.* **2010**, *13*, 338–349. [CrossRef]
30. Cherfan, M.; Vallée, A.; Kab, S.; Salameh, P.; Goldberg, M.; Zins, M.; Blacher, J. Unhealthy Behavior and Risk of Hypertension: The CONSTANCES Population-Based Cohort. *J. Hypertens.* **2019**, *37*, 2180–2189. [CrossRef]
31. Yang, Y.; Yu, D.; Piao, W.; Huang, K.; Zhao, L. Nutrient-Derived Beneficial for Blood Pressure Dietary Pattern Associated with Hypertension Prevention and Control: Based on China Nutrition and Health Surveillance 2015–2017. *Nutrients* **2022**, *14*, 3108. [CrossRef]
32. Fung, T.T.; Chiuve, S.E.; McCullough, M.L.; Rexrode, K.M.; Logroscino, G.; Hu, F.B. Adherence to a DASH-Style Diet and Risk of Coronary Heart Disease and Stroke in Women. *Arch. Intern. Med.* **2008**, *168*, 713–720. [CrossRef]
33. Bazzano, L.A.; Green, T.; Harrison, T.N.; Reynolds, K. Dietary Approaches to Prevent Hypertension. *Curr. Hypertens. Rep.* **2013**, *15*, 694–702. [CrossRef]
34. Tarray, R.; Saleem, S.; Afroze, D.; Yousuf, I.; Gulnar, A.; Laway, B.; Verma, S. Role of Insulin Resistance in Essential Hypertension. *Cardiovasc. Endocrinol.* **2014**, *3*, 129–133. [CrossRef]
35. Rodrigo, R.; González, J.; Paoletto, F. The Role of Oxidative Stress in the Pathophysiology of Hypertension. *Hypertens. Res.* **2011**, *34*, 431–440. [CrossRef]
36. Guyton, A.C. Blood Pressure Control—Special Role of the Kidneys and Body Fluids. *Science* **1991**, *252*, 1813–1816. [CrossRef] [PubMed]

37. Boegehold, M.A. The Effect of High Salt Intake on Endothelial Function: Reduced Vascular Nitric Oxide in the Absence of Hypertension. *J. Vasc. Res.* **2013**, *50*, 458–467. [CrossRef] [PubMed]
38. Oh, Y.S. Arterial Stiffness and Hypertension. *Clin. Hypertens.* **2018**, *24*, 17. [CrossRef]
39. Salvi, P.; Giannattasio, C.; Parati, G. High Sodium Intake and Arterial Stiffness. *J. Hypertens.* **2018**, *36*, 754–758. [CrossRef]
40. García-Ortiz, L.; Recio-Rodríguez, J.I.; Rodríguez-Sánchez, E.; Patino-Alonso, M.C.; Agudo-Conde, C.; Rodríguez-Martín, C.; Castaño-Sánchez, C.; Runkle, I.; Gómez-Marcos, M.A. Sodium and Potassium Intake Present a J-Shaped Relationship with Arterial Stiffness and Carotid Intima-Media Thickness. *Atherosclerosis* **2012**, *225*, 497–503. [CrossRef]
41. Kaess, B.M.; Rong, J.; Larson, M.G.; Hamburg, N.M.; Vita, J.A.; Levy, D.; Benjamin, E.J.; Vasan, R.S.; Mitchell, G.F. Aortic Stiffness, Blood Pressure Progression, and Incident Hypertension. *JAMA* **2012**, *308*, 875–881. [CrossRef]
42. Reynolds, A.N.; Akerman, A.; Kumar, S.; Diep Pham, H.T.; Coffey, S.; Mann, J. Dietary Fibre in Hypertension and Cardiovascular Disease Management: Systematic Review and Meta-Analyses. *BMC Med.* **2022**, *20*, 139. [CrossRef]
43. Fu, L.; Zhang, G.; Qian, S.; Zhang, Q.; Tan, M. Associations between Dietary Fiber Intake and Cardiovascular Risk Factors: An Umbrella Review of Meta-Analyses of Randomized Controlled Trials. *Front. Nutr.* **2022**, *9*, 972399. [CrossRef]
44. Aljuraiban, G.S.; Griep, L.M.O.; Chan, Q.; Daviglus, M.L.; Stamler, J.; van Horn, L.; Elliott, P.; Frost, G.S. Total, Insoluble and Soluble Dietary Fibre Intake in Relation to Blood Pressure: The INTERMAP Study. *Br. J. Nutr.* **2015**, *114*, 1480–1486. [CrossRef]
45. Guo, R.; Li, N.; Yang, R.; Liao, X.Y.; Zhang, Y.; Zhu, B.F.; Zhao, Q.; Chen, L.; Zhang, Y.G.; Lei, Y. Effects of the Modified DASH Diet on Adults with Elevated Blood Pressure or Hypertension: A Systematic Review and Meta-Analysis. *Front Nutr.* **2021**, *8*, 621. [CrossRef]
46. Soltani, S.; Arablou, T.; Jayedi, A.; Salehi-Abargouei, A. Adherence to the Dietary Approaches to Stop Hypertension (DASH) Diet in Relation to All-Cause and Cause-Specific Mortality: A Systematic Review and Dose-Response Meta-Analysis of Prospective Cohort Studies. *Nutr. J.* **2020**, *19*, 37. [CrossRef] [PubMed]
47. Murad, M.H.; Asi, N.; Alsawas, M.; Alahdab, F. New Evidence Pyramid. *BMJ Evid. Based Med.* **2016**, *21*, 125–127. [CrossRef] [PubMed]
48. Ramachandran, K.M.; Tsokos, C.P. *Mathematical Statistics with Applications in R*, 3rd ed.; Ramachandran, K.M., Tsokos, C.P., Eds.; Elsevier: Amsterdam, The Netherlands, 2020; ISBN 9780128178157.
49. Dobrescu, A.I.; Nussbaumer-Streit, B.; Klerings, I.; Wagner, G.; Persad, E.; Sommer, I.; Herkner, H.; Gartlehner, G. Restricting Evidence Syntheses of Interventions to English-Language Publications Is a Viable Methodological Shortcut for Most Medical Topics: A Systematic Review. *J. Clin. Epidemiol.* **2021**, *137*, 209–217. [CrossRef]
50. Morrison, A.; Polisena, J.; Husereau, D.; Moulton, K.; Clark, M.; Fiander, M.; Mierzwinski-Urban, M.; Clifford, T.; Hutton, B.; Rabb, D. The Effect of English-Language Restriction on Systematic Review-Based Meta-Analyses: A Systematic Review of Empirical Studies. *Int. J. Technol. Assess. Health Care* **2012**, *28*, 138–144. [CrossRef]

Disclaimer/Publisher's Note: The statements, opinions and data contained in all publications are solely those of the individual author(s) and contributor(s) and not of MDPI and/or the editor(s). MDPI and/or the editor(s) disclaim responsibility for any injury to people or property resulting from any ideas, methods, instructions or products referred to in the content.

 nutrients

Article

Effects of a Web-Based Weight Loss Program on the Healthy Eating Index-NVS in Adults with Overweight or Obesity and the Association with Dietary, Anthropometric and Cardiometabolic Variables: A Randomized Controlled Clinical Trial

Jan Kohl [1,*], Judith Brame [1], Pascal Hauff [1], Ramona Wurst [1], Matthias Sehlbrede [2], Urs Alexander Fichtner [2], Christoph Armbruster [2], Iris Tinsel [2], Phillip Maiwald [2], Erik Farin-Glattacker [2], Reinhard Fuchs [1], Albert Gollhofer [1] and Daniel König [1,3,4]

1 Department of Sport and Sport Science, University of Freiburg, 79117 Freiburg, Germany
2 Section of Health Care Research and Rehabilitation Research (SEVERA), Medical Center-University of Freiburg, Faculty of Medicine, University of Freiburg, 79106 Freiburg, Germany
3 Department of Sport Science, Institute for Nutrition, Exercise and Health, University of Vienna, 1150 Vienna, Austria
4 Department of Nutritional Sciences, Institute for Nutrition, Exercise and Health, University of Vienna, 1090 Vienna, Austria
* Correspondence: jan.kohl@sport.uni-freiburg.de

Citation: Kohl, J.; Brame, J.; Hauff, P.; Wurst, R.; Sehlbrede, M.; Fichtner, U.A.; Armbruster, C.; Tinsel, I.; Maiwald, P.; Farin-Glattacker, E.; et al. Effects of a Web-Based Weight Loss Program on the Healthy Eating Index-NVS in Adults with Overweight or Obesity and the Association with Dietary, Anthropometric and Cardiometabolic Variables: A Randomized Controlled Clinical Trial. *Nutrients* **2023**, *15*, 7. https://doi.org/10.3390/nu15010007

Academic Editor: Giuseppe Della Pepa

Received: 28 November 2022
Revised: 16 December 2022
Accepted: 16 December 2022
Published: 20 December 2022

Copyright: © 2022 by the authors. Licensee MDPI, Basel, Switzerland. This article is an open access article distributed under the terms and conditions of the Creative Commons Attribution (CC BY) license (https://creativecommons.org/licenses/by/4.0/).

Abstract: This randomized, controlled clinical trial examined the impact of a web-based weight loss intervention on diet quality. Furthermore, it was investigated whether corresponding changes in diet quality were associated with changes in measures of cardiovascular risk profile. Individuals with a body mass index (BMI) of 27.5 to 34.9 kg/m^2 and an age of 18 to 65 y were assigned to either an interactive and fully automated web-based weight loss program focusing on dietary energy density (intervention) or a non-interactive web-based weight loss program (control). Examinations were performed at baseline (t0), after the 12-week web-based intervention (t1), and after an additional 6 (t2) and 12 months (t3). Based on a dietary record, the Healthy Eating Index-NVS (HEI-NVS) was calculated and analyzed using a robust linear mixed model. In addition, bootstrapped correlations were performed independently of study group to examine associations between change in HEI-NVS and change in dietary, anthropometric, and cardiometabolic variables. A total of $n = 153$ participants with a mean BMI of 30.71 kg/m^2 (SD 2.13) and an average age of 48.92 y (SD 11.17) were included in the study. HEI-NVS improved significantly in the intervention group from baseline (t0) to t2 ($p = 0.003$) and to t3 ($p = 0.037$), whereby the course was significantly different up to t2 ($p = 0.013$) and not significantly different up to t3 ($p = 0.054$) compared to the control group. Independent of study group, there was a significant negative association between change in HEI-NVS and dietary energy density. A higher total score in HEI-NVS did not correlate with improvements in cardiovascular risk profile. The interactive and fully automated web-based weight loss program improved diet quality. Independent of study group, changes in HEI-NVS correlated with changes in energy density, but there was no association between improvements in HEI-NVS and improvements in cardiovascular risk profile.

Keywords: dietary quality; weight loss; cardiometabolic risk factors; body composition; dietary energy density; web-based intervention; fully automated; overweight; obesity

1. Introduction

A high-quality diet, together with adequate physical activity, is a cornerstone in the prevention and treatment of overweight or obesity and related non-communicable diseases such as cardiovascular disease, cancer or type 2 diabetes [1]. Thus, on the one hand, the increasing sedentary lifestyle has crucial negative health effects [2]. On the other hand, the excessive consumption of high energy density foods rich in sugar and fat, such as sweets, high-fat meat or cheese, has been shown to promote higher energy intake, weight gain and the risk of overweight and obesity [3–6]. It has been suggested that lowering dietary energy density, in addition to reducing dietary quantity [7,8], may also have a positive impact on diet quality [9–11]. A central role of a high-quality and low-energy-dense diet is the consumption of fruits and vegetables, which, with their low energy density and high amount of fiber, can make an important contribution to satiety and the supply of essential micronutrients [8,12]. In this regard, a high intake of fruits and vegetables is associated with a lower risk of cardiovascular disease, cancer, and all-cause mortality in observational studies [13,14]. Nevertheless, only few individuals meet the national recommendations for their intake. In Germany, according to the National Nutrition Survey II (NVS II), 87.4% of those examined fall below the 400 g recommendation for daily vegetable intake of the German Nutrition Society (DGE) and 59% of the people did not reach the recommendation of 250 g fruit per day [15,16].

A balanced diet according to the recommendations of the DGE [17], the Dietary Guidelines for Americans [18] or the Mediterranean diet [19] with sufficient intake of fruit, vegetables, protein dairy products, fish and whole grains as well as moderation in spreadable fats, alcohol and meat should prevent overweight and non-communicable diseases [1,20,21]. Due to the multidimensionality of health aspects in nutrition, the Healthy Eating Index (HEI) is a useful tool to evaluate nutrition in its entirety. At the same time, a HEI allows assessment of whether dietary patterns are consistent with dietary recommendations. With regard to the DGE recommendations, the HEI-NVS [17,22] was developed based on the HEI-1995 [23] and HEI-EPIC [24], to assess whether dietary patterns are consistent with national recommendations. Studies on this are relevant because, in addition to weight loss in overweight and obesity, a healthy diet allows direct beneficial effects, through bioactive substances such as unsaturated fats, phytochemicals, fiber or micronutrients [25–27] and should therefore be additionally evaluated as part of a nutritional intervention.

Dietary quality indices such as the HEI are commonly used in cross-sectional and observational studies to examine associations between scores and various health outcomes or parameters. However, in order to evaluate the effectiveness of a dietary intervention, its use is also becoming increasingly important in intervention studies to assess the quality of nutrition over the course of an intervention [28,29]. While the association between diet quality indices and anthropometric or cardiometabolic variables has been well studied in cross-sectional studies [30–32] as well as the health outcomes in long-term cohort studies [1], the health-related effects of diet quality changes have been less well studied in comparatively short-term intervention studies. Limited evidence suggests that behavioral weight loss interventions can improve diet quality [29]. Whether changes in a diet quality index are associated with changes in cardiometabolic, anthropometric or other dietary variables during an intervention is sparsely studied.

The results of the NVS II showed that adherence to national nutrition recommendations in Germany, surveyed using the HEI-NVS, was low. On average, men had 67 and women 69 out of a possible 110 points [33]. Experience has demonstrated that interventions with a high reach and long duration are needed to support long-term behavior change [34]. Web-based interventions could provide a cost-effective alternative to face-to-face programs and meet outreach and accessibility requirements [35–37], but according to recent reports on fitness trends from the American College of Sports Medicine, the popularity of such web-based interventions is still comparatively low [38,39]. Increased technical capabilities and a more robust scientific base mean that web-based interventions are becoming more interactive and tailored, which improves the effectiveness [40]. Emerging evidence suggests that web-based interventions can promote healthy eating behavior [41–44], while studies failed to show significant effects during a web-based weight loss intervention [45]. Therefore, further research is needed to examine the interplay of web-based interventions for weight loss on diet quality and whether changes in dietary quality are associated with changes in other nutritional or physiological variables.

This intervention study aims to evaluate the effects of two different web-based weight loss programs on diet quality assessed by the HEI-NVS. The intervention group received a fully automated and interactive web-based weight loss program focusing on dietary energy density, while the control group was exposed to a non-interactive web-based weight loss program (informative website) which addressed the same topics. We hypothesize that the interactive web-based weight loss program would have a statistically significant positive effect on HEI-NVS and that this effect would be significantly greater than in non-interactive web-based weight loss program. Furthermore, this analysis will examine whether, independent of group allocation, changes in HEI-NVS are associated with changes in energy density, energy intake, anthropometric or cardiometabolic variables. This manuscript was prepared according to the CONSORT-EHEALTH checklist (File S1).

2. Materials and Methods

2.1. Study Design

This randomized controlled clinical trial contained two groups running in parallel, which were allocated by permuted block randomization in a 1-to-1 ratio [46]. Participants in the online questionnaire study, which examined German-language web-based weight loss programs independent of location, were eligible to participate in this clinical substudy if they resided in southwestern Germany (postal code beginning with 79). In this clinical study, participants were invited to the Department of Sport and Sports Science and underwent medical examinations. In addition to medical variables, the dietary and physical activity behavior of the participants was investigated. All variables were collected at baseline (t0), after the 12-week web-based intervention (t1), and after additional 6 (t2) and 12-month (t3) follow-up.

2.2. Participants and Recruitment

Participants in the online questionnaire study were notified of the opportunity to take part in the clinical substudy after enrollment if they provided the place of residence with postal code beginning with 79 [46,47]. For the clinical trial, people of any gender, age between 18 and 65 years, and body mass index from 27.5 to 34.9 kg/m^2 were eligible to participate. Reasons for exclusion were breastfeeding or pregnancy as well as health problems or diseases. If existing health problems did not speak against participation in the program, this had to be certified with a medical certificate. Since the registration for the online questionnaire study as well as the registration for the clinical substudy took place online, appropriate computer skills were necessary. These were also required to use the web-based programs.

Various print and online media were used to recruit subjects for the clinical trial. Before study participants of southwestern Germany could register in the clinical substudy, they received the information on the study and had to provide written informed consent. After successful registration, randomization in the clinical substudy took place. In the subsequent telephone screening, potential study participants were again informed in detail about the study and the inclusion and exclusion criteria were reviewed. If the inclusion and exclusion criteria were not violated, an appointment was made for the baseline examination (t0). There, the final review of the criteria took place. Study participants received the Fitbit Charge 3 activity tracker (Fitbit, Inc.; San Francisco, CA, USA) as an incentive, which served as a measurement tool to record physical activity. Detailed information on participants and recruitment can be found in the study protocol [46].

2.3. Intervention

The intervention group's interactive web-based program was divided into three sections. In the first section, diet could be documented and appropriate feedback was provided in terms of energy density, energy intake, and macronutrients. In addition, various activities could be selected to pursue personal goals. These activities were aimed at making the diet healthier, reducing energy intake and increasing physical activity. If an activity was selected, it was scheduled accordingly in the personal calendar.

The second section included evidence-based information on energy density, weight loss, and healthy eating. Topics were divided into articles and some were part of weekly tasks. The third area included personal statistics and feedback. Through this section, the own progress could be monitored.

In contrast, the control group received a non-interactive web-based program that covered the same topics by means of pure knowledge transfer. The information was divided into short articles, but there was no algorithm-controlled feedback and the diet could not be recorded. A detailed description of the intervention can be found in the study protocol [46].

2.4. Outcome

A seven-day dietary record, which was to be maintained at all measurement time points, was used to calculate HEI-NVS [46]. The HEI-NVS consists of 10 components (fruits, vegetables, grains, milk, meat, fish, eggs, spreadable fats, beverages and alcohol) and allows a maximum of 110 points. The components and scoring standards of the HEI-NVS can be found in Table S1 based on Wittig and Hoffmann [22]. While a maximum score of 15 is possible for the fruits and vegetables components, 10 points are possible for the remaining 8 components. Dietary records were obtained using the nutritional software NutriGuide Plus (Version 4.8, Nutri-Science GmbH; Freiburg, Germany). The logged food entries were assigned to the different components according to the logic of the HEI-NVS and the score for each component was calculated. The total HEI-NVS score was calculated from the sum of the component scores. According to the logic of the HEI, a higher score represents a healthier diet and a diet closer to the recommendations of the DGE. Thus, the full HEI-NVS score of 110 corresponds to a dietary behavior within the recommendations of the DGE.

In addition to dietary data, anthropometric and cardiometabolic variables were collected [46]. Body weight, fat mass, fat free mass and body height were analyzed with the validated bioelectrical impedance analysis scale Seca mBCA 515 [48–50] and the stadiometer Seca 274 (Seca GmbH & Co. KG; Hamburg, Germany). In addition, the waist circumference was measured with the Seca 201 (Seca GmbH & Co. KG; Hamburg, Germany) measuring tape. Study staff took standardized measurements between the lowest rib and the iliac crest [46]. Blood pressure was assessed using a clinically validated device (Boso Medicus Exclusive, BOSCH + SOHN GmbH & Co. KG; Jungingen, Germany). Furthermore, blood samples were taken and analyzed by the Clotten Medical Care Center (MVZ) in Freiburg. Blood lipids (total cholesterol, HDL cholesterol, LDL cholesterol), blood glucose (fasting blood glucose, HbA1c) and other variables not relevant in this analysis were collected. A detailed description of the measurements and outcomes has been described elsewhere [46].

2.5. Sample Size, Randomization and Blinding

Sample size was calculated using the primary outcome of body weight with an estimated dropout rate of 15%. The calculation resulted in a sample size of 150 (75 + 75) participants. Participants were randomly assigned to the two interventions in a 1:1 allocation ratio using permuted block randomization with variable blocks. The allocation sequence was generated by the Section of Health Care Research and Rehabilitation Research of the University Freiburg (SEVERA) using RITA software (version 1.50, University of Lübeck; Lübeck, Germany). Allocation of subjects was automated upon their registration for the study.

Because subjects could figure out their allocated program based on study information, blinding of subjects was not possible. Outcome assessors were blinded, whereas data analysts were not. Details on sample size calculation, randomization and blinding have been described elsewhere [46,47].

2.6. Data Analysis

All statistical analyses were performed using R (Version 4.1.3) and RStudio (Version 2021.09.1). Two analyses were conducted. First, a per protocol analysis (PP analysis) was performed with the complete cases (cases without missing values). Second, an intention-to-treat analysis (ITT analysis) was carried out using multiple imputation (in total 50 imputations), with all randomized cases included. For multiple imputation, the R package micemd [51] was used. In both analyses, the total HEI-NVS score was analyzed with a robust linear mixed model and a significance level at 0.05. The R packages lme4 [52] and robustlmm [53] were used for this purpose. Visualization of the descriptive results was performed using the R package ggplot2 [54]. Because the results of the PP and ITT analyses were comparable, only the ITT analysis is presented here, which is the primary analysis according to the CONSORT-EHEALTH checklist.

To examine the association between changes in HEI-NVS and changes in dietary, anthropometric and cardiometabolic variables independent of group, bootstrapped Pearson correlation was performed and a biased corrected 95% confidence interval calculated using the R package boot [55]. For this purpose, the difference of the corresponding variables of t1 minus t0 as well as t3 minus t0 was calculated. A bootstrap sample size of 5000 was used to investigate associations in the imputed data (ITT analysis). Due to the imputed data set, all of the n = 153 subjects could be included and bootstrapping was performed with replacement to draw with n = 153 cases.

3. Results

3.1. Recruitment, Drop-Outs and Baseline Characteristics

From January to July 2020, n = 257 interested individuals registered for the clinical substudy (Figure 1). Registered individuals were contacted by phone and checked for inclusion and exclusion criteria. If these criteria were not violated, the individuals were invited to the baseline examination, where a final screening of the criteria took place. After these screenings, n = 153 individuals successfully completed the baseline examination. During the course of the study, n = 35 (22.9%) dropouts were observed. In both groups, dietary data were available for n = 52 subjects each across all measurement time points. These n = 104 subjects could therefore be included in the PP analysis of the dietary data. The baseline characteristics of participants in the two study groups were consistently similar and are shown in Table 1.

Table 1. Baseline (t0) characteristics of the study participants [1].

Variables	All (n = 153)	Intervention (n = 78)	Control (n = 75)
Age [years]	48.92 (11.17)	49.12 (11.36)	48.72 (11.05)
Sex			
Male [n]	44 (28.8%)	20 (25.7%)	24 (32.0%)
Female [n]	109 (71.2%)	58 (74.3%)	51 (68.0%)
Body weight [kg]	88.39 (10.65)	88.42 (10.15)	88.36 (11.21)
Body height [m]	1.69 (0.08)	1.69 (0.07)	1.70 (0.08)
BMI [kg/m^2]	30.71 (2.13)	30.88 (2.2)	30.54 (2.05)

[1] Data are presented as mean (SD) and frequencies (%).

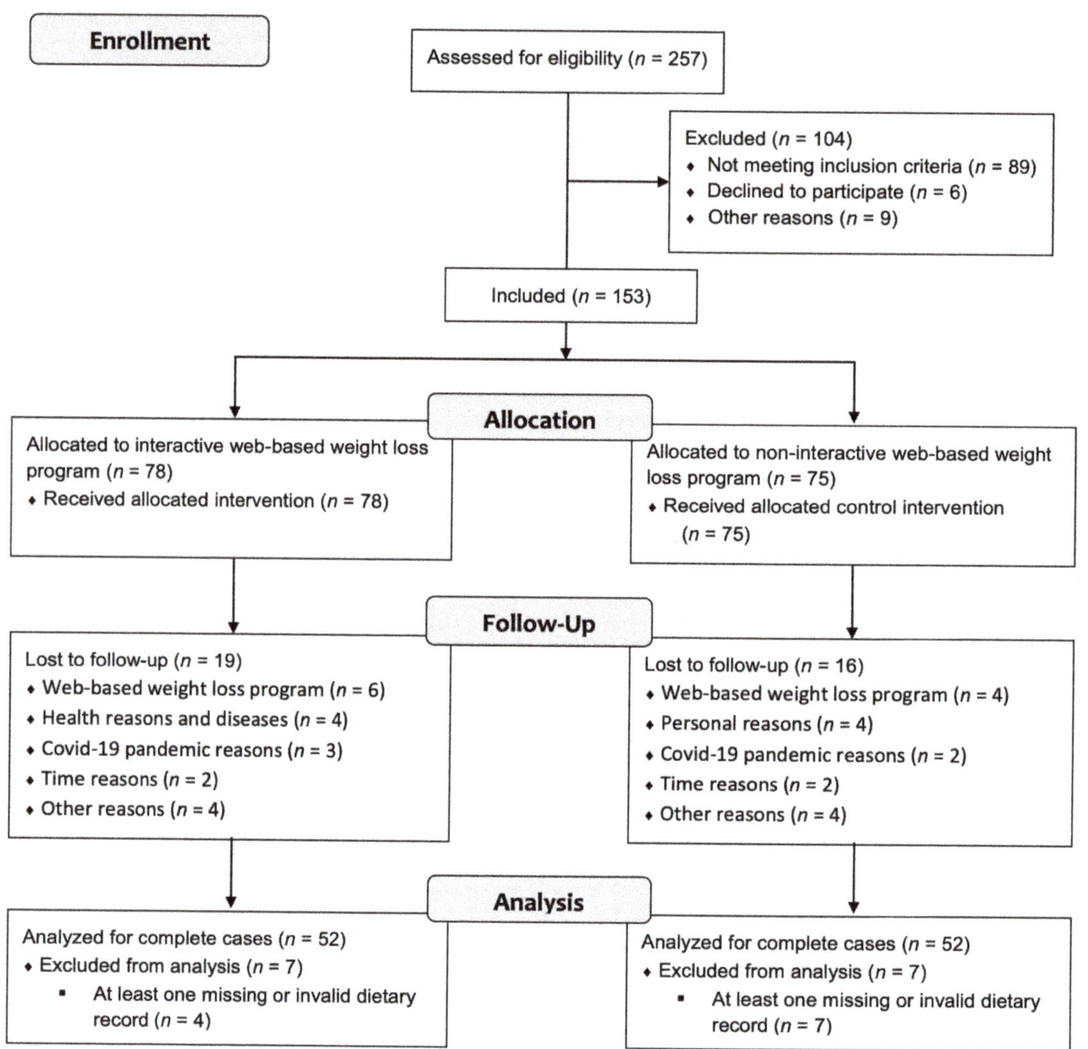

Figure 1. CONSORT flow chart depicting participant recruitment and drop-outs.

3.2. *Effects of Web-Based Weight Loss Programs on HEI-NVS*

The total HEI-NVS score increased significantly within the intervention group from baseline (t0) to t2 and t3, but not to t1 (Figure 2 and Table 2. Descriptively, the intervention group improved over the course of the study, particularly in the vegetables, fish and meat component (Table 3). The increase of the total score corresponded to a small effect from baseline to every measurement timepoint (Table 4). Compared with the control group, which deteriorated slightly from a descriptive point of view (Table 4), the statistical analysis showed a significantly different course from baseline to t2 and non-significant to t1 and t3 (Table 2). An analysis on the effects of the web-based programs on weight loss can be found elsewhere [56].

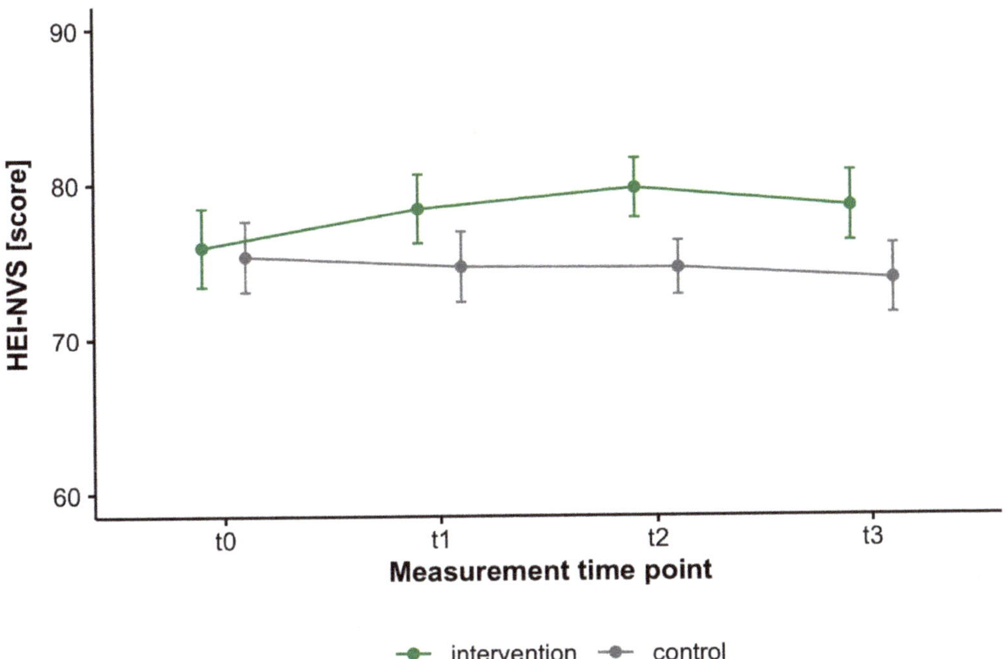

Figure 2. Mean and 95% confidence interval of the HEI-NVS for intervention (n = 78) and control (n = 75) (ITT analysis).

Table 2. Results of the robust linear mixed model of the HEI-NVS (ITT analysis) [1].

Predictors	HEI-NVS	*p*
Intercept	77.33 (2.66)	<0.001
Time		
t0–t1	5.23 (2.81)	0.063
t0–t2	9.06 (3.04)	0.003
t0–t3	5.90 (2.82)	0.037
Group (Control)	−1.10 (1.69)	0.513
Time * group (Control)		
t0–t1	−2.84 (1.79)	0.113
t0–t2	−4.96 (2.00)	0.013
t0–t3	−3.50 (1.81)	0.054

[1] Unstandardized regression coefficients with standard errors in parentheses.

Table 3. Descriptive statistics of components of the HEI-NVS (ITT analysis) [1].

Group	t0	t1	t2	t3
		Vegetables [score], max. 15 points		
Intervention	6.24 (3.41)	7.71 (4.06)	7.36 (3.01)	7.60 (3.51)
Control	6.15 (3.15)	6.67 (3.10)	5.72 (2.54)	6.57 (3.08)
		Fruits [score], max. 15 points		
Intervention	7.95 (4.37)	7.89 (4.36)	7.75 (3.96)	7.34 (4.41)
Control	7.31 (4.69)	7.20 (4.63)	6.42 (3.61)	6.77 (3.99)

Table 3. *Cont.*

Group	t0	t1	t2	t3
		Grains [score], max. 10 points		
Intervention	6.67 (2.35)	6.42 (2.19)	6.98 (2.05)	6.56 (2.05)
Control	6.71 (2.33)	6.35 (2.04)	7.00 (1.94)	6.84 (2.30)
		Dairy [score], max. 10 points		
Intervention	7.08 (2.15)	6.80 (1.95)	7.14 (1.75)	7.01 (2.05)
Control	7.15 (1.82)	7.33 (1.57)	7.27 (1.66)	7.14 (1.55)
		Fish [score], max. 10 points		
Intervention	3.29 (3.90)	3.87 (3.85)	4.07 (3.36)	4.15 (3.39)
Control	4.31 (3.93)	3.18 (3.53)	3.52 (3.24)	3.13 (3.33)
		Beverages [score], max. 10 points		
Intervention	8.93 (2.08)	8.97 (2.17)	8.87 (2.00)	8.75 (2.09)
Control	8.19 (2.52)	8.10 (2.84)	8.24 (2.48)	8.22 (2.59)
		Eggs [score], max. 10 points		
Intervention	8.79 (1.92)	8.91 (1.82)	8.89 (1.62)	8.84 (1.75)
Control	8.72 (2.10)	8.31 (2.31)	8.90 (1.75)	8.62 (2.10)
		Spreadable fats [score], max. 10 points		
Intervention	9.83 (0.82)	9.92 (0.43)	9.94 (0.37)	9.95 (0.22)
Control	9.76 (1.03)	9.91 (0.57)	9.93 (0.24)	9.90 (0.41)
		Alcohol [score], max. 10 points		
Intervention	9.23 (1.79)	9.29 (1.54)	9.57 (1.17)	9.45 (1.17)
Control	9.04 (1.93)	9.11 (1.75)	9.30 (1.57)	9.18 (1.65)
		Meat [score], max. 10 points		
Intervention	7.96 (2.43)	8.52 (1.98)	8.85 (1.51)	8.15 (1.94)
Control	7.95 (2.31)	8.13 (2.14)	8.08 (1.94)	7.83 (2.26)

[1] Intervention (n = 78) and control (n = 75) over four measurement time points (t0: 0 months, t1: 3 months, t2: 6 months after t1, t3: 12 months after t1). Data are presented as mean values (SD).

Table 4. Effect sizes (Cohen's d) with 95% confidence interval of the HEI-NVS (ITT analysis) [1].

Group	t0–t1	t0–t2	t0–t3
		HEI-NVS	
Intervention	0.24 [−0.08, 0.55]	0.38 [0.06, 0.70]	0.24 [−0.07, 0.56]
Control	−0.07 [−0.39, 0.25]	−0.09 [−0.41, 0.23]	−0.15 [−0.48, 0.17]

[1] Interpretation: $|d|$ = 0.2: small effect, $|d|$ = 0.5: medium effect, $|d|$ = 0.8: large effect.

3.3. Associations between HEI-NVS and Dietary, Anthropometric and Cardiometabolic Variables

The relationship between changes in HEI-NVS and other variables over the study period is presented in Table 5. Descriptive statistics of all variables used in the analysis can be found in Table S2. The changes in total HEI-NVS score correlated inversely with changes in energy density independent of group. Moreover, a weak positive correlation was observed between the change in HEI-NVS and the change in fasting blood glucose as well as a weak negative correlation with fat-free mass from t0 to t3. Apart from these findings, the analysis showed no further correlations between changes in HEI-NVS and other cardiovascular risk profile variables.

Table 5. Association between changes in HEI-NVS and changes in dietary, anthropometric, and cardiometabolic variables (ITT analysis) [1].

	Δt0–t1		Δt0–t3	
Variables	Correlation Coefficient	95% Confidence Interval	Correlation Coefficient	95% Confidence Interval
Energy density	−0.228 *	−0.359, −0.097	−0.312 *	−0.451, −0.165
Energy intake	0.089	−0.079, 0.256	0.076	−0.098, 0.247
Body weight	−0.052	−0.203, 0.122	−0.070	−0.235, 0.101

Table 5. *Cont.*

Variables	Δt0–t1		Δt0–t3	
	Correlation Coefficient	95% Confidence Interval	Correlation Coefficient	95% Confidence Interval
Waist circumference	0.068	−0.086, 0.216	−0.014	−0.203, 0.189
Fat mass	0.040	−0.103, 0.226	0.042	−0.111, 0.193
Fat free mass	−0.045	−0.209, 0.148	−0.190 *	−0.334, −0.041
Total cholesterol	−0.041	−0.185, 0.127	−0.018	−0.177, 0.133
HDL-cholesterol	−0.013	−0.165, 0.159	0.011	−0.163, 0.189
LDL-cholesterol	−0.087	−0.228, 0.065	0.001	−0.137, 0.151
Fasting blood glucose	−0.116	−0.258, 0.056	0.161 *	0.038, 0.275
HbA1c	−0.083	−0.217, 0.059	−0.055	−0.166, 0.081
Systolic blood pressure	0.104	−0.057, 0.264	−0.042	−0.221, 0.125
Diastolic blood pressure	0.176	−0.009, 0.365	−0.117	−0.297, 0.033

[1] Bootstrapped Pearson correlation with biased corrected confidence interval. * Statistically significant correlation.

4. Discussion

The main finding of the present study was that a fully automated and interactive web-based health program focusing on the dietary energy density improved the total HEI-NVS score, thereby shifting participants' diets toward the DGE dietary recommendations. Descriptively, these improvements were primarily due to improvement in the vegetables, fish, and meat components and resulted in a small effect in HEI-NVS from baseline to all three measurement timepoints. Compared to the non-interactive web-based weight loss program, however, there was only a significant advantage after 6-month follow-up (t0 to t2) and a non-significant difference after the 12-week intervention (t0 to t1) and after 12-month follow-up (t0 to t3). According to the meta-analysis published by Beleigoli and colleagues [45], none of the investigated web-based weight loss intervention demonstrated a significant advantage in diet quality over the control group. It should be noted that the included studies used very different instruments to measure dietary quality [45], which makes comparability difficult. Another meta-analysis on individuals with non-communicable diseases [44] showed benefits of eHealth interventions on healthy eating behaviors. The definition of healthy eating behaviors used in the studies included in this analysis had little overlap with diet quality. Thus, the outcomes used for inclusion were energy intake, macronutrient composition, and core food groups such as fruits or vegetables. While core food groups are often part of diet quality indices, diet quality is otherwise distinct from diet quantity and can only be represented to a limited extent, if at all, by macronutrient composition.

A recent systematic review demonstrated that weight loss interventions can improve diet quality as measured by a HEI [29]. Included studies covered in-person and mobile health interventions, which mostly resulted in an improvement between 4 to 7 points. In our study, mean improvements in the intervention group from baseline ranged from about 2.5 to 3.8 points, depending on the time of measurement. Thus, the improvement in this study tended to be lower than in the analysis by Cheng and colleagues [29]. However, it should be noted that the values cannot be directly compared because the review includes only studies using the U.S. versions HEI-2005, HEI-2010, and HEI-2015. These differ from each other and also from the HEI-NVS partly in components and evaluation system.

Besides the effect on the HEI-NVS of the web-based weight loss program focusing on reducing energy density, it was another important finding of the study that the change in HEI-NVS showed a weak to moderate inverse correlation with the change in energy density. An inverse relationship between energy density and diet quality has already been demonstrated in cross-sectional studies in various countries, e.g., Spain, Iran or Brazil [9–11], but to the best of our knowledge not yet in an intervention study. In contrast, improvements in HEI-NVS were not correlated with improvements in cardiovascular risk profile variables or changes in energy intake [57]. In contrast, the German National Nutrition Survey II, a representative cross-sectional study, found a positive association between HEI-NVS and BMI in women and in the 5th quintile in men [57]. In this longitudinal weight loss study, however, this was not confirmed with regard to body weight. Based on the absolute reference values of the HEI-NVS, it seems plausible that a higher score is associated with a higher energy intake and therefore a higher body weight or BMI. Also, with regard to the correlation between change in energy intake and HEI-NVS, this was not confirmed in this analysis.

Interestingly, this analysis revealed a weak positive correlation of change in HEI-NVS with change in fasting blood glucose as well as a weak negative correlation with change in fat-free mass. These results are surprising because it is assumed that improvement in a diet quality index is associated with better outcomes in cardiometabolic variables. There are several possible explanations for the lack of association between HEI-NVS and improvements in cardiometabolic variables. As previously reported, the observed effects on cardiometabolic variables by the two interventions were small. The effects on cardiometabolic variables have already been studied in the context of weight loss in people with overweight or obesity [58] and are consistent with the effects found in this study. Weight loss may already explain these effects, and the influence of diet quality may be insignificant and minor in the context of a weight loss intervention.

Furthermore, the construction of a diet quality index influences whether it correlates with diet quantity and thus with body weight and possibly other anthropometric variables. A negative association was shown between the change in HEI-NVS and the change in energy density, but not the change in energy intake. It is plausible that energy intake/dietary quantity and dietary quality may overlap if a diet quality index is not constructed appropriately. It is in fact possible that the HEI-NVS does not reflect diet quality independent of quantity. Due to the lack of reference to dietary quantity, as for example in the Healthy Eating Index-2015, a complete delineation to dietary quantity is compromised, as shown by the positive association between BMI and total score in women and partly in men [57]. Thus, the HEI-NVS measures compliance with the absolute amounts recommended in the German dietary guidelines rather than independent diet quality based on component reference values, which may tend to result in higher total scores if energy intake is high [57].

The following limitations must be considered when interpreting the results. First, the reference values of the HEI-NVS and the DGE, respectively, are based on systematic literature research and thus represent an important aspect for the preventive and therapeutic effects of nutrition in addition to the dietary quantity. However, considering other dietary indices and current findings in nutritional science, it is clear that important components of dietary quality such as sodium intake or carbohydrate quality such as intake of whole grains, refined grains, or added sugars are missing. The components and construction of the HEI-NVS may ultimately be responsible for the fact that improvements in HEI-NVS were not associated with improvements in cardiometabolic variables or, on the contrary, are partly even associated with negative effects on fasting blood glucose and fat-free mass. Thus, a differently constructed diet quality index might yield a different result.

Second, the reliability of self-reported data such as dietary records is limited. Recent data with reference data based on the doubly labeled water method suggests frequent underreporting, especially among people with overweight and obesity [59,60].

Third, compared to the real-world setting, both study groups may have been additionally motivated due to the activity tracker received as incentives as well as the free medical examinations provided by the study. In addition, complete blinding was not possible as subjects were likely to recognize their assigned program based on study information received in advance.

Fourth, the COVID-19 pandemic may also have influenced both groups in their dietary behavior. Because the COVID-19 pandemic and its limitations took a seasonal course, changes in dietary behavior are difficult to differentiate from seasonal changes and influence of the COVID-19 pandemic. The consumption of numerous food groups such as fruits, vegetables, or cereals, as well as energy intake, follows a seasonal pattern [61].

Simultaneously, an influence on dietary behavior could also be observed due to COVID-19 restrictions [62].

The elaborate implementation of seven-day dietary protocols is a strength of the present study. In addition, numerous anthropometric and cardiometabolic variables could be collected in a standardized manner in the clinical study. Therefore, this randomized controlled clinical trial provides more detailed insights into diet quality during web-based weight loss interventions and the association with dietary, anthropometric, and cardiometabolic variables.

5. Conclusions

Although the effect on HEI-NVS was small in the intervention group, this study demonstrated that a fully automated web-based weight loss intervention with focus on dietary energy density improved compliance with the national dietary recommendation. This result is relevant for all people aiming to reduce their body weight and eat healthier at the same time, but do not have access to personal care. Furthermore, the change in HEI-NVS showed an inverse correlation with the change in dietary energy density. Improvements in HEI-NVS were not associated with improvements in anthropometric and cardiometabolic variables. Interestingly, improvements in HEI-NVS were associated with only weak unfavorable effects on fat-free mass and fasting blood glucose. One might speculate that a diet quality index addressing other components, such as whole grains or salt, would have found more beneficial relationships. In the future, more intervention studies should address the association of diet quality and cardiovascular risk factors to examine the short-term effects of diet quality. This would provide a better understanding of the health effects of diet quality or related indices.

Supplementary Materials: The following supporting information can be downloaded at: https://www.mdpi.com/article/10.3390/nu15010007/s1, Table S1: Components and scoring standards of the HEI-NVS; Table S2: Descriptive statistics of variables used for correlation independent of study group. File S1: CONSORT-EHEALTH Checklist.

Author Contributions: Conceptualization, J.K., J.B., R.W., M.S., U.A.F., C.A., I.T., P.M., E.F.-G., R.F., A.G. and D.K.; methodology, J.K., J.B. and D.K.; formal analysis, J.K.; investigation, J.K., J.B. and P.H.; resources, J.B. and D.K.; data curation, J.K., J.B. and P.H.; writing—original draft preparation, J.K.; writing—review and editing, J.K., J.B., P.H., R.W., M.S., U.A.F., C.A., I.T., P.M., E.F.-G., R.F., A.G. and D.K.; visualization, J.K.; project administration, J.K., J.B. and D.K.; funding acquisition, D.K. All authors have read and agreed to the published version of the manuscript.

Funding: This research was funded by the Techniker Krankenkasse (German Health Insurance Company). The APC was funded by the Open Access Publication Fund of the University of Freiburg.

Institutional Review Board Statement: The study was conducted in accordance with the Declaration of Helsinki, and approved by the ethics commission of Albert-Ludwigs-University, Medical Center, Freiburg (vote no. 237/19) on 25 July 2019. The study was registered in the German Clinical Trials Register before the commencement of the intervention (DRKS00020249, https://www.drks.de), which is approved by the World Health Organization. Due to the results of two clinical pilot studies (vote no. 409/18, DRKS00016512), minor changes were made to the study protocol and positively assessed by the ethics commission (date of approval: 22 October 2019, protocol version: amendment 01).

Informed Consent Statement: Informed consent was obtained from all subjects involved in the study.

Data Availability Statement: The data of this study are the property of the Department of Sport and Sport Science of the University of Freiburg. The results of the clinical trials are presented to the scientific community and the trial sponsor as aggregated research data. Data on individual subjects cannot be reconstructed. No further data of the clinical trials are released to the scientific community or third parties.

Acknowledgments: We would like to thank all study participants as well as the study staff. Furthermore, we would like to thank Irina Kopman and her team at Vilua Healthcare GmbH for the extensive IT support. Finally, we would like to thank the Techniker Krankenkasse, represented by Nicole Knaack, Kerstin Hofreuter-Gätgens and Dagmar Köppel, for funding this study.

Conflicts of Interest: J.K., J.B., P.H. and D.K. report funding by the Techniker Krankenkasse for clinical trial design, implementation, and scientific evaluation. R.W., R.F., M.S., I.T., E.F., C.A., U.A.F. and P.M. report funding by the Techniker Krankenkasse for the design, implementation, and scientific evaluation of the online trials. A.G. declares no conflict of interest. The funders had no role in the design of the study; in the collection, analyses, or interpretation of data; in the writing of the manuscript; or in the decision to publish the results.

References

1. Schwingshackl, L.; Bogensberger, B.; Hoffmann, G. Diet Quality as Assessed by the Healthy Eating Index, Alternate Healthy Eating Index, Dietary Approaches to Stop Hypertension Score, and Health Outcomes: An Updated Systematic Review and Meta-Analysis of Cohort Studies. *J. Acad. Nutr. Diet.* **2017**, *118*, 74–100.e11. [CrossRef]
2. Bull, F.C.; Al-Ansari, S.S.; Biddle, S.; Borodulin, K.; Buman, M.P.; Cardon, G.; Carty, C.; Chaput, J.-P.; Chastin, S.; Chou, R.; et al. World Health Organization 2020 guidelines on physical activity and sedentary behaviour. *Br. J. Sports Med.* **2020**, *54*, 1451–1462. [CrossRef]
3. World Health Organization. Obesity and Overweight. Available online: http://www.who.int/news-room/fact-sheets/detail/obesity-and-overweight (accessed on 3 August 2022).
4. Pérez-Escamilla, R.; Obbagy, J.E.; Altman, J.M.; Essery, E.V.; McGrane, M.M.; Wong, Y.P.; Spahn, J.M.; Williams, C.L. Dietary Energy Density and Body Weight in Adults and Children: A Systematic Review. *J. Acad. Nutr. Diet.* **2012**, *112*, 671–684. [CrossRef]
5. Rouhani, M.H.; Haghighatdoost, F.; Surkan, P.J.; Azadbakht, L. Associations between dietary energy density and obesity: A systematic review and meta-analysis of observational studies. *Nutrition* **2016**, *32*, 1037–1047. [CrossRef]
6. Stelmach-Mardas, M.; Rodacki, T.; Dobrowolska-Iwanek, J.; Brzozowska, A.; Walkowiak, J.; Wojtanowska-Krosniak, A.; Zagrodzki, P.; Bechthold, A.; Mardas, M.; Boeing, H. Link between Food Energy Density and Body Weight Changes in Obese Adults. *Nutrients* **2016**, *8*, 229. [CrossRef]
7. Robinson, E.; Khuttan, M.; McFarland-Lesser, I.; Patel, Z.; Jones, A. Calorie reformulation: A systematic review and meta-analysis examining the effect of manipulating food energy density on daily energy intake. *Int. J. Behav. Nutr. Phys. Act.* **2022**, *19*, 48. [CrossRef]
8. Rouhani, M.H.; Surkan, P.J.; Azadbakht, L. The effect of preload/meal energy density on energy intake in a subsequent meal: A systematic review and meta-analysis. *Eat. Behav.* **2017**, *26*, 6–15. [CrossRef]
9. Schröder, H.; Mendez, M.A.; Gomez, S.F.; Fíto, M.; Ribas, L.; Aranceta, J.; Serra-Majem, L. Energy density, diet quality, and central body fat in a nationwide survey of young Spaniards. *Nutrition* **2013**, *29*, 1350–1355. [CrossRef]
10. Mendes, A.; Pereira, J.L.; Fisberg, R.; Marchioni, D.M.L. Dietary energy density was associated with diet quality in Brazilian adults and older adults. *Appetite* **2015**, *97*, 120–126. [CrossRef]
11. Azadbakht, L.; Haghighatdoost, F.; Esmaillzadeh, A. Dietary energy density is inversely associated with the diet quality indices among Iranian young adults. *J. Nutr. Sci. Vitaminol.* **2012**, *58*, 29–35. [CrossRef]
12. Smethers, A.D.; Rolls, B.J. Dietary Management of Obesity: Cornerstones of Healthy Eating Patterns. *Med. Clin. N. Am.* **2018**, *102*, 107–124. [CrossRef]
13. Aune, D.; Giovannucci, E.; Boffetta, P.; Fadnes, L.T.; Keum, N.; Norat, T.; Greenwood, D.C.; Riboli, E.; Vatten, L.J.; Tonstad, S. Fruit and vegetable intake and the risk of cardiovascular disease, total cancer and all-cause mortality-a systematic review and dose-response meta-analysis of prospective studies. *Int. J. Epidemiol.* **2017**, *46*, 1029–1056. [CrossRef] [PubMed]
14. Angelino, D.; Godos, J.; Ghelfi, F.; Tieri, M.; Titta, L.; Lafranconi, A.; Marventano, S.; Alonzo, E.; Gambera, A.; Sciacca, S.; et al. Fruit and vegetable consumption and health outcomes: An umbrella review of observational studies. *Int. J. Food Sci. Nutr.* **2019**, *70*, 652–667. [CrossRef]
15. Heuer, T.; Krems, C.; Moon, K.; Brombach, C.; Hoffmann, I. Food consumption of adults in Germany: Results of the German National Nutrition Survey II based on diet history interviews. *Br. J. Nutr.* **2015**, *113*, 1603–1614. [CrossRef]
16. MRI. German National Nutrition Survey II 2005–2007. 2008. Available online: https://www.mri.bund.de/fileadmin/MRI/Institute/EV/NVSII_Abschlussbericht_Teil_2.pdf (accessed on 3 August 2022).
17. Deutsche Gesellschaft für Ernährung (DGE). *Food-Related Recommendations*, 1st ed.; DGE: Bonn, Germany, 2014.
18. US Department of Agriculture; US Department of Health and Human Services. *Dietary Guidelines for Americans, 2020–2025*, 9th ed.; 2020. Available online: https://www.dietaryguidelines.gov/sites/default/files/2020-12/Dietary_Guidelines_for_Americans_2020-2025.pdf (accessed on 10 November 2022).
19. Bach-Faig, A.; Berry, E.M.; Lairon, D.; Reguant, J.; Trichopoulou, A.; Dernini, S.; Medina, F.X.; Battino, M.; Belahsen, R.; Miranda, G.; et al. Mediterranean diet pyramid today. Science and cultural updates. *Public Health Nutr.* **2011**, *14*, 2274–2284. [CrossRef]
20. Schwingshackl, L.; Hoffmann, G.; Iqbal, K.; Schwedhelm, C.; Boeing, H. Food groups and intermediate disease markers: A systematic review and network meta-analysis of randomized trials. *Am. J. Clin. Nutr.* **2018**, *108*, 576–586. [CrossRef]

21. Schlesinger, S.; Neuenschwander, M.; Schwedhelm, C.; Hoffmann, G.; Bechthold, A.; Boeing, H.; Schwingshackl, L. Food Groups and Risk of Overweight, Obesity, and Weight Gain: A Systematic Review and Dose-Response Meta-Analysis of Prospective Studies. *Adv. Nutr. Int. Rev. J.* **2019**, *10*, 205–218. [CrossRef]
22. Wittig, F.; Heuer, T.; Claupein, E.; Pfau, C.; Cordts, A.; Schulze, B.; Padilla Bravo, C.A.; Spiller, A. (Eds.) *Auswertung der Daten der Nationalen Verzehrsstudie II (NVS II): Eine Integrierte Verhaltens- und Lebensbasierte Analyse des Bio-Konsums*; Rubner-Institut: Karlsruhe, Germany; Georg-August-University Göttingen: Göttingen, Germany, 2010; pp. 51–68.
23. Kennedy, E.T.; Ohls, J.; Carlson, S.; Fleming, K. The Healthy Eating Index: Design and applications. *J. Am. Diet. Assoc.* **1995**, *95*, 1103–1108. [CrossRef]
24. Rüsten, A.V.; Illner, A.K.; Boeing, H.; Flothkötter, M. Evaluation of food intake based on a "Healthy Eating Index" [HEI-EPIC]. *Ernahr. Umsch.* **2009**, *56*, 450–456.
25. Hooper, L.; Martin, N.; Jimoh, O.F.; Kirk, C.; Foster, E.; Abdelhamid, A.S. Reduction in saturated fat intake for cardiovascular disease. *Cochrane Database Syst. Rev.* **2020**, *5*, CD011737. [CrossRef]
26. Speer, H.; D'Cunha, N.M.; Botek, M.; McKune, A.J.; Sergi, D.; Georgousopoulou, E.; Mellor, D.D.; Naumovski, N. The Effects of Dietary Polyphenols on Circulating Cardiovascular Disease Biomarkers and Iron Status: A Systematic Review. *Nutr. Metab. Insights* **2019**, *12*, 1178638819882739. [CrossRef]
27. Reynolds, A.; Mann, J.; Cummings, J.; Winter, N.; Mete, E.; Te Morenga, L. Carbohydrate quality and human health: A series of systematic reviews and meta-analyses. *Lancet* **2019**, *393*, 434–445. [CrossRef]
28. Brauer, P.; Royall, D.; Rodrigues, A. Use of the Healthy Eating Index in Intervention Studies for Cardiometabolic Risk Conditions: A Systematic Review. *Adv. Nutr. Int. Rev. J.* **2021**, *12*, 1317–1331. [CrossRef]
29. Cheng, J.; Liang, H.-W.; Klem, M.L.; Costacou, T.; Burke, L.E. Healthy Eating Index Diet Quality in Randomized Weight Loss Trials: A Systematic Review. *J. Acad. Nutr. Diet.* **2023**, *123*, 117–143. [CrossRef]
30. Sullivan, V.K.; Petersen, K.S.; Fulgoni, V.L., 3rd; Eren, F.; Cassens, M.E.; Bunczek, M.T.; Kris-Etherton, P.M. Greater Scores for Dietary Fat and Grain Quality Components Underlie Higher Total Healthy Eating Index-2015 Scores, While Whole Fruits, Seafood, and Plant Proteins Are Most Favorably Associated with Cardiometabolic Health in US Adults. *Curr. Dev. Nutr.* **2021**, *5*, nzab015. [CrossRef]
31. Artegoitia, V.M.; Krishnan, S.; Bonnel, E.L.; Stephensen, C.B.; Keim, N.L.; Newman, J.W. Healthy eating index patterns in adults by sex and age predict cardiometabolic risk factors in a cross-sectional study. *BMC Nutr.* **2021**, *7*, 30. [CrossRef]
32. Miller, V.; Webb, P.; Micha, R.; Mozaffarian, D. Defining diet quality: A synthesis of dietary quality metrics and their validity for the double burden of malnutrition. *Lancet Planet. Health* **2020**, *4*, e352–e370. [CrossRef]
33. Gose, M.; Krems, C.; Heuer, T.; Hoffmann, I. Trends in food consumption and nutrient intake in Germany between 2006 and 2012: Results of the German National Nutrition Monitoring (NEMONIT). *Br. J. Nutr.* **2016**, *115*, 1498–1507. [CrossRef]
34. Castelnuovo, G.; Pietrabissa, G.; Manzoni, G.M.; Corti, S.; Ceccarini, M.; Borrello, M.; Giusti, E.M.; Novelli, M.; Cattivelli, R.; Middleton, N.A.; et al. Chronic care management of globesity: Promoting healthier lifestyles in traditional and mHealth based settings. *Front. Psychol.* **2015**, *6*, 1557. [CrossRef]
35. Bennett, G.G.; Glasgow, R.E. The Delivery of Public Health Interventions via the Internet: Actualizing Their Potential. *Annu. Rev. Public Health* **2009**, *30*, 273–292. [CrossRef]
36. Murray, E. Web-Based Interventions for Behavior Change and Self-Management: Potential, Pitfalls, and Progress. *Medicine 2.0* **2012**, *1*, e1741. [CrossRef]
37. Brown, V.; Tran, H.; Downing, K.L.; Hesketh, K.D.; Moodie, M. A systematic review of economic evaluations of web-based or telephone-delivered interventions for preventing overweight and obesity and/or improving obesity-related behaviors. *Obes. Rev.* **2021**, *22*, e13227. [CrossRef]
38. Kercher, V.M.; Kercher, K.; Bennion, T.; Levy, P.; Alexander, C.; Amaral, P.C.; Li, Y.-M.; Han, J.; Liu, Y.; Wang, R.; et al. 2022 Fitness Trends from Around the Globe. *ACSM'S Health Fit. J.* **2022**, *26*, 21–37. [CrossRef]
39. Batrakoulis, A. European survey of fitness trends for 2020. *ACSM'S Health Fit. J.* **2019**, *23*, 28–35. [CrossRef]
40. Sorgente, A.; Pietrabissa, G.; Manzoni, G.M.; Rethlefsen, M.; Simpson, S.; Perona, S.; Rossi, A.; Cattivelli, R.; Innamorati, M.; Jackson, J.B.; et al. Web-Based Interventions for Weight Loss or Weight Loss Maintenance in Overweight and Obese People: A Systematic Review of Systematic Reviews. *J. Med. Internet Res.* **2017**, *19*, e229. [CrossRef]
41. Fiedler, J.; Eckert, T.; Wunsch, K.; Woll, A. Key facets to build up eHealth and mHealth interventions to enhance physical activity, sedentary behavior and nutrition in healthy subjects—An umbrella review. *BMC Public Health* **2020**, *20*, 1605. [CrossRef]
42. Vandelanotte, C.; Müller, A.M.; Short, C.E.; Hingle, M.; Nathan, N.; Williams, S.L.; Lopez, M.L.; Parekh, S.; Maher, C.A. Past, Present, and Future of eHealth and mHealth Research to Improve Physical Activity and Dietary Behaviors. *J. Nutr. Educ. Behav.* **2016**, *48*, 219–228.e1. [CrossRef]
43. Dening, J.; Islam, S.M.S.; George, E.; Maddison, R. Web-Based Interventions for Dietary Behavior in Adults With Type 2 Diabetes: Systematic Review of Randomized Controlled Trials. *J. Med. Internet Res.* **2020**, *22*, e16437. [CrossRef]
44. Duan, Y.; Shang, B.; Liang, W.; Du, G.; Yang, M.; Rhodes, R.E. Effects of eHealth-Based Multiple Health Behavior Change Interventions on Physical Activity, Healthy Diet, and Weight in People With Noncommunicable Diseases: Systematic Review and Meta-analysis. *J. Med. Internet Res.* **2021**, *23*, e23786. [CrossRef]

45. Beleigoli, A.M.; Andrade, A.Q.; Cançado, A.G.; Paulo, M.N.; Diniz, M.D.F.H.; Ribeiro, A.L. Web-Based Digital Health Interventions for Weight Loss and Lifestyle Habit Changes in Overweight and Obese Adults: Systematic Review and Meta-Analysis. *J. Med. Internet Res.* **2019**, *21*, e298. [CrossRef]
46. Brame, J.; Kohl, J.; Wurst, R.; Fuchs, R.; Tinsel, I.; Maiwald, P.; Fichtner, U.; Armbruster, C.; Bischoff, M.; Farin-Glattacker, E.; et al. Health Effects of a 12-Week Web-Based Lifestyle Intervention for Physically Inactive and Overweight or Obese Adults: Study Protocol of Two Randomized Controlled Clinical Trials. *Int. J. Environ. Res. Public Health* **2022**, *19*, 1393. [CrossRef]
47. Tinsel, I.; Metzner, G.; Schlett, C.; Sehlbrede, M.; Bischoff, M.; Anger, R.; Brame, J.; König, D.; Wurst, R.; Fuchs, R.; et al. Effectiveness of an interactive web-based health program for adults: A study protocol for three concurrent controlled-randomized trials (EVA-TK-Coach). *Trials* **2021**, *22*, 526. [CrossRef]
48. Jensen, B.; Braun, W.; Geisler, C.; Both, M.; Klückmann, K.; Müller, M.J.; Bosy-Westphal, A. Limitations of Fat-Free Mass for the Assessment of Muscle Mass in Obesity. *Obes. Facts* **2019**, *12*, 307–315. [CrossRef]
49. Bosy-Westphal, A.; Schautz, B.; Later, W.; Kehayias, J.J.; Gallagher, D.; Müller, M.J. What makes a BIA equation unique? Validity of eight-electrode multifrequency BIA to estimate body composition in a healthy adult population. *Eur. J. Clin. Nutr.* **2013**, *67*, S14–S21. [CrossRef]
50. Bosy-Westphal, A.; Jensen, B.; Braun, W.; Pourhassan, M.; Gallagher, D.; Müller, M.J. Quantification of whole-body and segmental skeletal muscle mass using phase-sensitive 8-electrode medical bioelectrical impedance devices. *Eur. J. Clin. Nutr.* **2017**, *71*, 1061–1067. [CrossRef]
51. Audigier, V.; Resche-Rigon, M. Micemd: Multiple Imputation by Chained Equations with Multilevel Data. 2021. Available online: https://cran.r-project.org/web/packages/micemd/index.html (accessed on 27 November 2022).
52. Bates, D.; Mächler, M.; Bolker, B.; Walker, S. Fitting Linear Mixed-Effects Models Using lme4. *J. Stat. Softw.* **2015**, *67*, 1–48. [CrossRef]
53. Koller, M. robustlmm: An *R* Package for Robust Estimation of Linear Mixed-Effects Models. *J. Stat. Softw.* **2016**, *75*, 1–24. [CrossRef]
54. Wickham, H. *ggplot2: Elegant Graphics for Data Analysis*; Springer: New York, NY, USA, 2016.
55. Canty, A.; Ripley, B. *Boot: Bootstrap R (S-Plus) Functions*; R Core Team: Vienna, Austria, 2021. Available online: https://cran.r-project.org/web/packages/boot/citation.html (accessed on 27 November 2022).
56. Kohl, J.; Brame, J.; Centner, C.; Wurst, R.; Fuchs, R.; Sehlbrede, M.; Tinsel, I.; Maiwald, P.; Fichtner, U.A.; Armbruster, C.; et al. Effects of a web-based lifestyle intervention in adults with overweight and obesity on weight loss and cardiometabolic risk factors: A randomized controlled clinical trial. *JMIR* 2022, *preprints*.
57. Moon, K.; Krems, C.; Heuer, T.; Roth, A.; Hoffmann, I. Predictors of BMI Vary along the BMI Range of German Adults–Results of the German National Nutrition Survey II. *Obes. Facts* **2017**, *10*, 38–49. [CrossRef]
58. Morris, E.; Jebb, S.; Oke, J.; Nickless, A.; Ahern, A.; Boyland, E.; Caterson, I.; Halford, J.; Hauner, H.; Aveyard, P. How effective is weight loss in reducing cardiometabolic risk? An observational analysis of two randomised controlled trials of community weight loss programmes. *Br. J. Gen. Pract.* **2021**, *71*, e312–e319. [CrossRef]
59. Burrows, T.L.; Ho, Y.Y.; Rollo, M.E.; Collins, C.E. Validity of Dietary Assessment Methods When Compared to the Method of Doubly Labeled Water: A Systematic Review in Adults. *Front. Endocrinol.* **2019**, *10*, 850. [CrossRef] [PubMed]
60. Park, Y.; Dodd, K.W.; Kipnis, V.; Thompson, F.E.; Potischman, N.; Schoeller, D.A.; Baer, D.J.; Midthune, D.; Troiano, R.; Bowles, H.; et al. Comparison of self-reported dietary intakes from the Automated Self-Administered 24-h recall, 4-d food records, and food-frequency questionnaires against recovery biomarkers. *Am. J. Clin. Nutr.* **2018**, *107*, 80–93. [CrossRef] [PubMed]
61. Stelmach-Mardas, M.; Kleiser, C.; Uzhova, I.; Peñalvo, J.L.; La Torre, G.; Palys, W.; Lojko, D.; Nimptsch, K.; Suwalska, A.; Linseisen, J.; et al. Seasonality of food groups and total energy intake: A systematic review and meta-analysis. *Eur. J. Clin. Nutr.* **2016**, *70*, 700–708. [CrossRef] [PubMed]
62. Buckland, N.J.; Swinnerton, L.F.; Ng, K.; Price, M.; Wilkinson, L.L.; Myers, A.; Dalton, M. Susceptibility to increased high energy dense sweet and savoury food intake in response to the COVID-19 lockdown: The role of craving control and acceptance coping strategies. *Appetite* **2020**, *158*, 105017. [CrossRef] [PubMed]

Disclaimer/Publisher's Note: The statements, opinions and data contained in all publications are solely those of the individual author(s) and contributor(s) and not of MDPI and/or the editor(s). MDPI and/or the editor(s) disclaim responsibility for any injury to people or property resulting from any ideas, methods, instructions or products referred to in the content.

MDPI
St. Alban-Anlage 66
4052 Basel
Switzerland
www.mdpi.com

Nutrients Editorial Office
E-mail: nutrients@mdpi.com
www.mdpi.com/journal/nutrients

Disclaimer/Publisher's Note: The statements, opinions and data contained in all publications are solely those of the individual author(s) and contributor(s) and not of MDPI and/or the editor(s). MDPI and/or the editor(s) disclaim responsibility for any injury to people or property resulting from any ideas, methods, instructions or products referred to in the content.

www.ingramcontent.com/pod-product-compliance
Lightning Source LLC
LaVergne TN
LVHW070653170726
843515LV00004B/396

* 9 7 8 3 0 3 6 5 8 7 7 4 5 *